The Selfish Meme

A Critical Reassessment

Culture is a unique and fascinating aspect of the human species. How did it emerge and how does it develop? Richard Dawkins has suggested that culture evolves and that memes are the cultural replicators, subject to variation and selection in just the same way as genes are in the biological world. In this sense human culture is the product of a mindless evolutionary algorithm. Does this imply, as some have argued, that we are mere meme machines and that the conscious self is an illusion?

Kate Distin's highly readable and accessible book extends and strengthens Dawkins's theory and presents for the first time a fully developed and workable concept of cultural DNA. She argues that culture's development can be seen both as the result of memetic evolution and as the product of human creativity. Memetic evolution is perfectly compatible with the view of humans as conscious and intelligent.

This book should find a wide readership amongst philosophers, psychologists and sociologists, and it will also interest many nonacademic readers.

Kate Distin is an independent scholar.

Contents

CAMBRIDGE UNIVERSITY PRESS
Cambridge, New York, Melbourne, Madrid, Cape Town, Singapore,
São Paulo, Delhi, Dubai, Tokyo

Cambridge University Press
32 Avenue of the Americas, New York, NY 10013-2473, USA

www.cambridge.org
Information on this title: www.cambridge.org/9780521606271

© Kate Distin 2005

This publication is in copyright. Subject to statutory exception
and to the provisions of relevant collective licensing agreements,
no reproduction of any part may take place without the written
permission of Cambridge University Press.

First published 2005
Reprinted 2005, 2006, 2007

A catalog record for this publication is available from the British Library

Library of Congress Cataloging in Publication data

Distin, Kate 1970–
The selfish meme : a critical reassessment / Kate Distin.
p. cm.
Includes bibliographical references and index.
ISBN 0-521-84452-5 – ISBN 0-521-60627-6 (pbk.)
1. Culture. 2. Social evolution. 3. Human evolution. I. Title.
HM621.D57 2004
306'.01-dc22 2004051882

ISBN 978-0-521-84452-9 Hardback
ISBN 978-0-521-60627-1 Paperback

Transferred to digital printing 2009

Cambridge University Press has no responsibility for the persistence or
accuracy of URLs for external or third-party Internet websites referred to in
this publication, and does not guarantee that any content on such websites is,
or will remain, accurate or appropriate. Information regarding prices, travel
timetables and other factual information given in this work are correct at
the time of first printing but Cambridge University Press does not guarantee
the accuracy of such information thereafter.

The Selfish Meme

A Critical Reassessment

KATE DISTIN

1

Introduction

Shortly after we were married, my husband made me a mandolin. The body is built from rosewood and the bridge hand carved from ebony. Wood can be bent if you heat it, but he had no bending iron – so he curved the sides by rocking them over some hair curling tongs, clamped to the kitchen table.

I had wanted a mandolin since I was a child – for almost as long as I had been playing the violin. The two instruments have the same intervals between their strings, and it seemed to me that it must be easier to rest something across your lap, plucking at notes whose positions were marked out for you by frets, than to contort the whole of your upper body into the violinist's masochistic stance, attempting simultaneously to create notes on a standard scale with your left hand and to tame two feet of bow with your right. I already understood what instructions the notes on a stave were trying to give my fingers, and had lately been charmed by the mandolin music of Vivaldi and Oysterband. (I was naïve, as it happens. The mandolin does have all these advantages, but it also – as the fingers of my left hand will testify – has strings like cheese wire.)

My husband found the design in a woodworking magazine, tucked in amongst the usual advertisements and feature articles. An engineer by training, he had inherited both skills and tools from his father and grandfather. When the plans let him down, he spent some time thinking about the physics of the processes involved, learnt a bit about concert pitch, and then calculated the appropriate fret spacings from first principles. We read up on the mandolin's origins: where was it first invented, what sorts of music had people played on it, and for how many years? We were drawn into a study of the history of music, and debated over late-night bottles of wine whether its conventions were discovered or invented. By a pleasing coincidence, the hot novel of the year was *Captain Corelli's Mandolin*.

My mandolin is thus the end product of a trail of culture that stretches back across centuries and continents. Its creation was dependent on woodworking techniques and tools, on the development of stringed instruments and musical conventions, on the physics and mathematics of sound, and on the modern world of magazine articles and advertisements. As it grew, we were pointed in the directions of its historical and geographical origins, and our attention was drawn to philosophical and scientific theories about its music. It has links to a vast range of cultural areas, all of which are more like icebergs than mountains, their manifest modern complexities resting on unseen millennia of previous human thought and activity.

Richard Dawkins has said that "most of what is unusual about man can be summed up in one word: 'culture'."[1] Culture is not humans' only distinguishing feature, but it is one unique and fascinating aspect of our species. In this context, "culture" is not intended to be either a description of a narrow range of purely artistic pursuits or a synonym for society. "A *society* refers to an actual group of people and how they order their social relations. A *culture*... refers to a body of socially transmitted information"[2] – the full spectrum of ideas, concepts and skills that is available to us in society. It includes science and mathematics, carpentry and engineering designs, literature and viticulture, systems of musical notation, advertisements and philosophical theories – in short, the collective product of human activities and thought.

How did this body of knowledge and methods emerge? How does it now continue to develop? This book defends the theory that culture *evolves*, and that *memes* provide the mechanism for that evolution.

"Evolution" is usually taken to apply only in the biological world, referring to the theory developed by Charles Darwin and others in the nineteenth century to account for the origin of species. In the twentieth century, Richard Dawkins and others pointed out that the core of Darwinian theory is actually rather sparse. Its essential elements are simply replication, variation and selection. If these requirements are met then evolution seems bound to happen. If organisms reproduce, passing their characteristics almost (but not always quite) accurately on to the next generation, and if their environment does not supply them with unlimited resources for their survival, then they will evolve: there will be a struggle for survival, and those organisms will be preserved whose traits are best fitted to the given environment. It is the business of science to investigate the actual pattern of development in our natural world, but at a more theoretical level Darwin's theory outlines a process that is inevitable once all of its elements are in place. Dawkins has suggested that this not only is true in biology but would also be the case in any other environment where all of those key elements were to be found – including culture.

The suggestion that evolution is not restricted to biology but may also take place in culture is appealing if hardly original. A metaphorical picture is often painted of ideas and theories "evolving" through time, but can it ever with justification be taken literally? That is, does Darwinism illustrate a process that can also be observed in culture? There are various versions of cultural evolutionary theory on the market, and this book explores what I see as the most compelling: the hypothesis that the units of cultural selection are elements, which Dawkins calls "memes", that share the important properties of genes.

The biggest danger for this hypothesis is the risk of its collapsing into the trivial assertion that some ideas survive whilst others disappear. Obviously cultures change, ideas spread and technology develops, but what do we gain by claiming that this is all due to memetic evolution? What does the meme hypothesis contribute to our understanding which other theories of cultural change do not?

One way of responding to this challenge is to take a very practical approach, and seek out areas of culture to which meme theory can fruitfully be applied. Most other books on memes have tended to follow this line, and have developed memetic explanations of phenomena such as religion, language and the size of the human brain. The best way of defending the meme hypothesis, from this perspective, is to show that it can provide useful accounts of developments in such key cultural areas. This is an approach in the best traditions of scientific experiment, using observation to confirm or falsify novel theories. Given a hypothesis about how culture develops, practical observations about what actually happens in human culture will surely provide a good method of testing its validity.

On the other hand, it is notoriously difficult to ensure that we take an objective view of the evidence when we are seeking to confirm a favoured hypothesis. Indeed, it is sometimes possible to present very different explanations of the same observations, each of which makes perfect sense from the perspective of a given hypothesis, but only one of which can be true. The history of science is littered with theories that once convinced the brightest of contemporary intellects, and our adversarial justice system also bears testimony to the potential for weaving different stories from the same body of evidence.

Behind that evidence, however, lies the truth, and there must surely be a more direct approach to its discovery. Rather than testing the practical implications of a novel theory, an alternative is to focus first on its underlying structure: to examine whether it *could* be true, is internally coherent and forms a solid basis for any empirical applications. Inevitably such investigations will have to take into account some observations of the phenomena that the theory purports to explain, but the focus will be on testing its structural foundations *before* trying to use it as a tool for scientific enquiry. It is this approach which I favour.

In the case of culture, for example, the question is not so much whether development in its various areas *can* be characterized as memetic as whether the meme hypothesis is *true*. According to Dawkins, culture evolves in the same way as biology – but in which ways, exactly, are the two processes "the same"? What would replication, variation or selection *be* in relation to culture? Is culture really made up of discrete units? To what extent can other concepts from genetics be transferred to memetics – concepts like *vehicle* and *phenotype, virus* and *allele*? Where are memes to be found, and what is the memetic equivalent of DNA?

Satisfactory answers to such questions will inevitably contribute to our understanding of cultural development. For example, if culture is a unique feature of humans, then meme theory should be able to explain what has enabled us to develop such a feature when nonhuman animals have not. Indeed, since other animals surely *do* pass on information and skills to each other, it should include an account of what is special about the "memes" that purportedly make up human culture. Supported by such theoretical investigations, it should of course be possible for meme theory to provide an account of how ideas change and develop in particular cultural areas like science and religion. Even more fundamentally, it should enable us to explore the relationship between memes and the human mind: do they create us, do we create them – or is there, as some would claim, no real difference between "us" and "them"?

This last question is obviously of huge significance for how we humans see ourselves. Some of the best-known names in the field – in particular Daniel Dennett and Susan Blackmore – believe that meme theory will completely overturn our traditional notions of responsibility, creativity and intentionality, just as many have taken the Darwinian revolution in biology to have overturned traditional notions of a creator God. On their view what we call our minds, with all their apparent powers of thought, decision making and invention, are actually parasitic meme complexes, our sense of control over which is illusory. If our mental and cultural lives are the results of a mindless evolutionary algorithm, they argue, then how can we claim an autonomous identity as independent "selves", with freedom and control over what goes on in those lives?

Despite the apparent power of this argument and the persuasiveness of its authors, my own conclusion – freely reached after many hours of genuinely creative thought and non-illusory choices – is that memetic evolution is quite consistent with a world of intentional, conscious and responsible free agents. And if it weren't, then common sense dictates that I should exercise my free will and reject meme theory in preference to dispensing with mind, conscience and autonomy. Fortunately, however, neither option is necessary, as this book will show.

2

The Meme Hypothesis

Richard Dawkins first proposed his version of cultural evolutionary theory in his 1976 book, *The Selfish Gene*. The main thrust of that book was a defence of the gene as the unit of biological selection and the organism as a "survival machine" for its genes. Towards the end, however, he added his view that culture also evolves and that "memes" are the units of *cultural* selection.

The key to Dawkins's idea is that Darwinian evolution is a particular instance of a process that we might also expect to find in other areas. It will be helpful, therefore, to begin with a swift review of Charles Darwin's theory of descent, before explaining how the meme hypothesis emerges from it. Having characterized Dawkins's own view of what has come to be termed "memetics", I then briefly defend its adoption against the alternative research programme of sociobiology. There is also in this chapter an important clarification of the relation between genes and memes. These introductory discussions provide history and context for the more detailed investigations of subsequent chapters.

Genetic Evolution

Natural Selection

In the early nineteenth century, the problem of the origin of species was so far from being solved that Darwin referred to it as the "mystery of mysteries". He worked on his own solution for more than two decades, until in 1859 *The Origin of Species* brought together a vast mass of previously isolated facts, all of which fitted into place when seen in the light of his theory of descent.

Darwin was famously inspired by Malthus to see that all organisms are engaged in a perpetual struggle for existence, due to the pressure of population on the available resources. Beginning with the facts that organisms in a species vary and that those variations are passed on to their offspring, he saw in addition that human beings have used this to their advantage by artificially selecting animals and plants with the most useful variations. He argued that, since organic variations useful to man have occurred, it seems likely that in thousands of generations some variations useful to each organism would also have occurred. If so, then because of the ongoing struggle for existence, any individual with such an advantage would have had the best chance of survival and procreation, and any injurious variation would lead to the destruction of its owner – with the result that those organisms are naturally selected which have the optimum fit to conditions of existence. Later, in *The Descent of Man* (1871), Darwin added that humans are subject to this evolutionary process just like any other animals. His view was that our unique mental features would one day be explicable by natural selection, which could also account for human social and ethical behaviour.

Genetics

Today, gene theory tells us that natural selection consists in the differential survival of *replicators* – things that make copies of themselves. In the struggle for existence, replicators with "longevity, fecundity, and copying fidelity"[1] will have a better chance of survival than others, and it is now widely accepted that in biology those replicators are genes. A preliminary sketch of gene theory, based largely on Richard Dawkins's account, will thus provide a useful backdrop to his meme hypothesis.

As a starting point, familiarity with a little vocabulary from the language of genetics would be helpful: jargon in its best sense is useful both as shorthand and as a conceptual tool. Although this is not the point to digress into the technical details of genetic replication – this book is after all written by a philosopher rather than a cellular biologist – gene theory does provide the background to memetics, and it will often prove fruitful to explore the analogy between the two. Thus: a *gene* "stores" the information that it replicates in deoxyribonucleic acid (*DNA*) – that is to say that the gene occupies a particular locus (place) on a *chromosome* (a structure within a cell nucleus), and the chromosome is composed of DNA; the gene may also have *alleles*, which are alternative forms of it in the population, occupying the same locus on that chromosome and controlling the same sorts of things as it does (e.g., eye colour) – its *phenotypic effects*.

The story goes that in the "primeval soup" the competition for resources and space meant that the ancestors of genes fared best if they had some means of protection. Over time, the protective mechanisms that they developed evolved into more complex "survival machines" – in our case, the human body. Although genes are the units of reproduction, their existence within these "survival machines", or vehicles, means that they are selected indirectly: their differential survival rates depend on their phenotypic effects. So long as they replicate accurately, their effects will also be passed on to the next generation, but when genes do not make exact copies their effects will vary too, and individuals will survive or be eliminated as a result of such (un)favourable variations. Continuous, gradual changes of this sort will result, through successive generations, in new species and types.

Another significant feature of evolution, as Dawkins sees it, is the *nature* of the replicators. Famously, he refers to genes as "selfish". By this he means that each behaves in such a way as to increase its own welfare at the expense of other genes in the gene pool. Successful adaptations will result in its longer life, say, or increased fecundity. He certainly does not mean to imply that genes are *consciously* seeking their own replication, but simply that they cannot survive if they are inefficient at self-replication.

Why Accept Gene Theory?

At the time that *The Origin* was published, Darwin's ideas were highly controversial in a way that they are not, amongst scientists, today. Nonetheless, even then emergent theses in palaeontology, biology and geology were all contributing to an intellectual climate which was more receptive to the novel idea that species might change over time, and Darwin capitalized on this by collecting a mass of evidence in support of his theory. When Mendel's gene hypothesis came to light, it seemed to be Darwin's final vindication, for it provided a mechanism for evolution.

Today the explanatory success of neo-Darwinism is undeniable. Seeing things from the genes' point of view allows us to explain all sorts of superficially puzzling phenomena. A well-known example is biological altruism, when members of a species behave in ways that benefit other individuals at their own expense: the individual's behaviour may be detrimental to his *own* survival, but it promotes the survival of close members of his species – members who (because they are relatives, or just very much like him) share many of his genes. Thus his behaviour is "altruistic" at the individual, but "selfish" at the genetic, level.*

This is a specific instance of explanatory success. In general, the point is that the survival of a species depends upon the preservation of its members' strengths. The existence of genes – units of transmission, to future generations, of the beneficial characteristics of the present generation – makes this possible. In particular, Mendel's theory of divisible and recombinable pairs of alleles provides the variation upon which selection can act. Not only via the recombination of genetic information, but also by its mutation (since genes' copying fidelity is not always exact), the gene pool varies, and selection ensures that advantageous variations are preserved. Hence, over enormous time spans, nature's immense variety can be explained. Such explanatory power justifies our acceptance of gene theory. Long before the identification of their physical basis in DNA there were very good reasons to believe in genes' existence, for they provide the basic material of selection.

The Meme Hypothesis

This, then, is the background to the meme hypothesis, which extrapolates from the Darwinian theory of biological evolution to apply the concept of selection more generally. As Dawkins puts it, "Darwinism is too big a theory to be confined to the narrow context of the gene":[2] its essential feature is the differential survival of replicators – *any* replicators. Whatever the type of replicator involved, Dawkins conjectures, its variation under conditions of restricted resources would lead to a form of evolution. There is a process at work here, whose function should in theory be unaffected by the medium upon which it is based. Just as the same sum can be performed by hand, on a calculator or on one of any number of computer spreadsheet programs, so Dawkins wonders whether the same evolutionary algorithm might be able to operate on a range of different units of selection.

His suggestion is attractive because it seems to strike a happy balance between the extremists who would bring everything under a pattern of development that mimics biological evolution, and those who prefer to restrict the concept purely to biology. Dawkins rejects such a

* I am talking here about biological altruism, rather than altruism as we might understand it in everyday speech: the genetic impartiality of religious doctrines such as "love thy neighbour" is arguably inexplicable at this level.

restriction as artificial, but nor does he tie himself to a particular pattern of development; rather, he extracts the significant features of evolutionary theory, and extends their domain of influence.

In particular he turns his attention to culture, which he sees as the distinctive feature of the human species. Cultural transmission does occur in species other than man, but not to the same vast extent. In humans alone, Dawkins hypothesizes another example of the process that Darwinism illustrates, in this case involving cultural replicators. These replicators he calls "memes", and he postulates a new form of selection such that "once the genes have provided their survival machines with brains that are capable of rapid imitation, the memes will automatically take over."[3] Dawkins defines a meme as "a unit of cultural inheritance, hypothesized as analogous to the particulate gene, and as naturally selected in virtue of its phenotypic consequences on its own survival and replication in the cultural environment".[4]

As examples of memes, he suggests ideas, catch-phrases, tunes (or snatches of tunes), fashions and skills.[†] As with genes, the constituents of success will be long life, fecundity and accuracy of replication; for individual copies, fecundity is the most important factor. The element of competition necessary for any selection to take place is introduced by the brain's limited attention: in order to dominate, a meme must distract the brain's attention from other memes. Success in this matter will depend upon the structure of the brain, as well as on the stability of the meme and its "penetrance in the cultural environment".[5] The latter will depend on psychological appeal, and according to Dawkins this means (as for genes) that coadapted complexes – that is, evolutionarily stable sets of memes – will occur. Selection will favour those memes capable of exploiting the current cultural environment, which obviously includes other memes also trying to be selected. As sets of memes cooperate, new ones will find it more difficult to penetrate the environment later: the complex provides protection against invasion. The compatibility rule will apply particularly in areas such as theories of science. For other types of memes different criteria will apply – "catchiness" for tunes, for example.

Dawkins emphasizes that their success will *not* depend on the (dis)advantages they produce for the genotypes that produced the brains

[†] In this section I provide an overview of meme theory as Dawkins himself has outlined it. Obviously there are areas here which other memeticists would find controversial, but as a preliminary sketch of the hypothesis I think it most appropriate to stick with its originator's views.

they inhabit. Memes, like genes, are selfish: their success depends on the advantages they confer on *themselves.* In the struggle for brains' attention they must in some way be "better" than their rivals, but this need have nothing to do with the effects they have on the genetic success of their possessors. Although the needs of genes and memes may often coincide (a meme will not last long if it causes its possessor to die before she can transmit it, for example), they may sometimes be in complete opposition: Dawkins uses the example of a meme for celibacy to illustrate this possibility.

He says that a meme will, like a gene, be successful "by proxy": via its phenotypic effects. The meme itself is a "unit of information residing in a brain",[6] and its phenotypic effects are the external consequences of that piece of information. Words, skills and music are "the outward and visible (audible, etc.) manifestations of the memes within the brain",[7] which are transmitted between individuals via their sense organs, leaving on the recipient's brain a (not necessarily exact) copy that it is free to transmit again. Dawkins notices that a gene's phenotypic effects take two forms: the use it makes of the cellular apparatus to make copies of itself; and the effects it has on the external world, which influence its survival chances. He says that memes also have two types of effect. The first is the use of their possessors' communication and imitation skills in order to replicate. The second, as for genes, consists in the effects they have on the world, which influence their survival chances. The success of this second type of effect will (for both kinds of replicator) depend on the current environment, a crucial part of which will be the existing pool of replicators.

The Gene-Meme Analogy

Notice that, although "memetic" evolution may for convenience be referred to as "analogous to" genetic evolution, this should not be taken to imply that memetics is theoretically dependent on genetics. Whilst it is true that in the chronological order of theoretical development memetic evolution has been inspired by the theory of genetic evolution, this is not the order of explanatory dependence. Rather, both are examples of a more abstract, generally applicable theory of the evolution of replicators under conditions of competition.

In fact the use of the term "analogy", in this context, deserves some closer attention. Usually this term implies that at least one side of the comparison is fairly well developed – we talk about electric "current", for instance, because water flow is familiar to us – and an analogy would not be of any use if this were not the case.

A different sort of scientific comparison can be made between subjects like gravity and electrostatic force: here a new student can easily see the similarity between the two laws below without being familiar with either field:

Newton's law of universal gravitation:
"The force between two masses is directly proportional to the product of their masses and inversely proportional to the square of their separation."

Coulomb's law:
"The force between two point charges is directly proportional to the product of their charges and inversely proportional to the square of their separation."

The relationship between memetics and genetics is best understood as a combination of these alternatives. A meme is not, strictly speaking, an analogue of a gene: rather, since both are replicators, a meme is a different *token* of the same *type* of entity as a gene. (The type-token distinction is a handy conceptual tool, of which I make fairly frequent use. A *token* is any "particular specimen of any general class. All these specimens may be described as the several tokens of that single *type*.")[8] Similarly, cultural evolution is a different example of the same type of process as neo-Darwinism, rather than a simple analogue of it. This means that the two processes have the same description at a sufficiently functional, abstract level.

Nonetheless, because we are already familiar with genetics, we can use it to illuminate memetics. In other words, although the particular details of biological evolution may not carry over into cultural evolution, it seems reasonable to exploit our knowledge of neo-Darwinism as a guide to what the essential elements of cultural evolution might be. This sort of comparison between two phenomena is far from unique in science, where it is quite common to find different tokens of the same type of process realized within different media (e.g., wave properties such as diffraction, interference and refraction may all be observed in water waves as well as in electromagnetic and sound waves). "Comparison of two examples is a good way to locate what is most important"; it helps in "pruning away content and leaving essentials".[9]

Why Accept the Meme Hypothesis?

Returning now to Dawkins's original hypothesis, the question arises what are the prima facie grounds for accepting it. This section explains why the memetic research programme looks promising, and for consistency it follows Charles Darwin's method of defence for his own evolutionary theory.

The starting point for Darwin's line of thought was that variations occur within the traits of a species, and that they are passed on to the offspring of the organisms that so vary. Variation seems to be a good starting place for a theory of cultural evolution, too. There are often marked differences amongst the knowledge and practices of those who would claim to have the same concept, skill or idea, and it is possible to trace the extent of such variation to the point where two people at either extreme of it would deny that they have the same idea at all.

As an example, take the ability to play the piano: some people are talented sight readers, others play from music but are hopeless sight readers, others struggle to read music but improvise well, and others play only by ear. Along this spectrum of players there will be people whose abilities are almost the same – those who can all read music but some of whom are better sight readers than others, for instance. If, though, we compare those at opposite ends of the spectrum, all of whom would describe themselves as pianists, then we can see that their abilities are so different that they might more accurately be classified into separate categories, such as concert pianists and jazz improvisers.

Are such variations passed on to offspring? Clearly, "offspring" does not here refer to biological but to cultural descendants – and it seems obvious that the variations *are* transmitted. If my piano teacher is a concert pianist then he will teach me to read music, with an emphasis on the repetitive practice of pieces that I have first sight-read; if he is a jazz pianist then he will teach me the techniques of improvisation and how to play by ear. Just as the variations that you inherit from your biological parents may develop differently in you, depending on the nature of your environment, so the variations that you acquire from your cultural predecessors (who might be your teachers, people whose books you have read, musicians whose style you have imitated, etc.) may develop differently in the context of your mind and environment. What matters from the point of view of evolution, however, is that those variations *are* replicated in you.

In culture as in biology, then, variations exist and are passed on to the next "generation". The next strand in Darwin's argument came from Malthus's theory of a population which increases much faster than its limited resources. From this, Darwin extracted the idea of the struggle for existence, which is another important structural feature of evolutionary theory. Does the cultural "population" also increase at a swifter rate than its resources?

The answer to this must depend on the definitions of a cultural population, and of the nature of its resources. The contenders for membership of the cultural population are controversial amongst memeticists, but I think that it can be stated without raising too many hackles that they are almost innumerable: ideas, concepts, skills, concertos, fashions, ways of building houses, farming methods, . . . These are all aspects of the cultural world that might potentially be passed on from one possessor of them to another. Therefore their resources must be human beings' attention. It seems obvious that culture and ideas develop and change at a much faster rate than that of biological evolution, and that the attention of each human brain is limited. In order to maintain some sort of grip on day-to-day reality we have to choose between the skills, theories and so on to which we might direct our efforts and which we might keep available in memory. It is just not possible for us to keep up with every available area of knowledge and skill. Thus it appears that the cultural population does increase faster than its resources.

From the transmission of variations and the struggle for existence, Darwin derived the idea of natural selection: in the struggle for the resources of a limited environment, those organisms with slightly advantageous variations will have a better chance of survival and replication, whereas those whose variations are at all deleterious will find their survival threatened. There is no great difficulty for cultural evolution with this stage of Darwin's argument, for it consists merely in deducing the consequences of the previous stages (though, as noted above, those consequences will be played out a far greater pace than is the case in biology). A form of selection must occur in minds and culture, and we should expect to see the preservation of those ideas and skills with the best fit to their environment, and the extinction of those without. At first glance, therefore, the meme hypothesis does hold some promise.

Sociobiology

This initial survey has raised the hope that a theory of cultural evolution can be developed along the same lines as the theory of biological evolution – but perhaps we should not forget that Darwin himself would have disagreed with such a project, since he believed that human behaviour can be attributed to just the same laws of descent as that of other species. Accordingly, some would argue that sociobiology is more appropriate than memetics as a means of studying human society. The aim of this discipline is, according to Edward Wilson – one of its founding fathers – "to show how social groups adapt to the environment by evolution".[10] That human society *is* greatly influenced by its genetic heritage Wilson does not doubt. The "accumulated evidence" for this, he describes as "decisive".

Now, there is nothing very interesting about the claim that broad, general aspects of social behaviour will, if advantageous, be selected: the point of society is to protect its members' genes and encourage their propagation, so behaviour or attitudes that tend to preserve social structures will (amongst social organisms) be favoured. The interesting question is what *level* of social detail our genes control – and I would argue that there is so much variation amongst cultures that it is highly implausible that many of the specific details should be genetically controlled.

The truth is, rather, that natural selection generally *obliterates* the heritable variation of the traits that it favours: as a result of being favoured, they become fixed throughout the population, and thereafter any variation amongst the relevant phenotypic effects must be explained environmentally.[11] Moreover, there is no reason to believe that the human brain of one or even two thousand years ago was dissimilar to ours; yet there is an enormous disparity between modern *culture* and that of previous millennia. This pace of change is much too rapid to be picked up at the level of genes, so if evolutionary theory is to be applied to such changes, then it will be more appropriate to bring it to bear on behaviour and the mind than on neural architecture and its genetic code.

Sociobiology asserts that the organic origin of the human capacity for culture ensures that "however it may work in detail, culture will usually enhance genetic fitness"[12] – and it is certainly true that our general capacity for culture could not have evolved had it not initially been adaptively advantageous: the early development of the mind and of culture must have provided us with a mechanism to ensure that more of the successful cultural traits were beneficial than were harmful to us, because we still exist. It seems likely, in any case, that an advanced capacity for learning *would* have increased fitness. On the other hand, this does not imply that each particular cultural trait will increase fitness, and must also be distinguished from the claim that there will still, today, be a general correlation between cultural habits' popularity and their helpfulness to us. Today the rate of cultural development is so great that most such developments will be *neutral* with respect to our biological survival.

Furthermore, a distinction should be drawn between the true fact that all human behaviour will fall within the potential range permitted by our genetic code, and the extrapolation from this to the invalid deduction that where there is adaptive behaviour there is always a genetic basis for it. Such a suggestion is undermined by the quite striking difference between the rates at which the physical and the cultural worlds change. Think, for example, of the development of the computer: its adaptive advantage is immeasurable, but it would be ridiculous to assert that the human genetic code has changed to accommodate it.

Clearly, the memetic project would be damaged if it turned out that the human mind *is* wholly innate. Conversely, sociobiology would be seriously undermined by the confirmation of the tabula rasa hypothesis that the mind is, at birth, a blank surface upon which experience writes. Since neither of these extremes is likely to be wholly true, the important question is, as mentioned, *where* we should draw the line between genetic and environmental (i.e., cultural) control over behaviour. The closer this line is to the "innate" extreme, the more significant will be some of the claims of sociobiology; the closer it is to the "cultural" extreme, the less plausible they will be. The discussion in this section implies that the development of the human mind is not so heavily genetically determined that the role left for culture is trivial, and therefore that cultural evolutionary theory will provide a complement, rather than a rival, to much of the account that sociobiology provides of human thinking and behaviour.

Towards an Adequate Theory of Cultural Evolution

This chapter has provided the beginnings of an argument to suggest that the evolutionary processes – replication, selection and variation – are present in culture, but an adequate theory of cultural evolution depends also on our ability to isolate the *aspects* of each process which are most significant in that realm. Memeticists claim that there are elements of culture which vary, are copied and selected, but this claim is – even when supported by observation and argument – much too vague to satisfy. We need to look deeper than this, investigating the *ways* in which cultural information is preserved; the *mechanisms* that enable such complex information to be replicated; the *causes* and *limits* of the variations that arise; the *factors* that influence selection amongst these variations.

Only then will it be possible to build a proper account of how culture evolves.

Chapters 3–6 examine each of these evolutionary processes in turn, demonstrating that all can be observed in culture as well as in biology, and making use of the genetic analogy to extract the key features of each.

3

Cultural DNA

The most basic element in evolution, whether biological or cultural, is replication. There are two steps involved in replication: the preservation of the information that is copied, and the means by which it is transmitted. This chapter asks in what form cultural information might be preserved.

In evolution of any form, what evolves is essentially information. Genes are a means of preserving biological information, and the format that they use is DNA. We know where to look for the units of biological selection (within organisms) and we know what form that information takes (DNA). In culture, however, things are not yet so obvious, and this is a real stumbling block for many who first encounter the meme hypothesis. It is all very well to suggest that culture "evolves" via memes, just as biology does via genes, but where exactly are these memes to be found and – most fundamentally – *what* are they?

The second half of this book looks in detail at the problem of memes' location, but this chapter concentrates on the issue of memes' underlying basis. Just as the course of genetic evolution has been shaped by its ultimate reliance on DNA, so the course of cultural evolution must ultimately be dependent on the nature of the information that is being selected. There was a time when Mendel's gene hypothesis was undermined by the absence of any real understanding of how heredity worked. It was not until Watson and Crick had revealed heredity's chemical basis, with the discovery of DNA's molecular structure, that genetics really took off. Similarly, until we can give an account of how cultural heredity works, the edifice of memetics will inevitably be weakened by this gap in its foundations. What, then, is the cultural equivalent of DNA?

Information and Its Effects

We can learn a lot about the nature of replicators by studying their most familiar incarnation, DNA. It is crucial to note that DNA preserves information between generations in a particular way. First, and most fundamentally, the information must be preserved in a form that allows it to be replicated. Secondly, since genes are selected via their phenotypic effects, the information must also be preserved in a form that enables its effects to be activated in a variety of contexts and situations. If the information cannot exert its phenotypic effects – or if the circumstances in which it can do so are too tightly restricted – then it will be unavailable for selection.

Since memes are replicators it is reasonable to expect that their content, too, must be preserved in a particular way. Like genes, the most fundamental requirement must be that their information is preserved in a form that allows it to be copied. Similarly, memes' information must also be preserved in such a form that it potentially has a phenotypic effect, via which it can be selected.

What does this mean in practice? Much information will have a severely restricted impact on the meme pool, owing to its limited effects on the world. The reasons for such limitations are varied. For example, the Spanish that I learned many years ago, for exam purposes, has now all but disappeared from my memory, since its *potential* effects (enabling me to communicate with other Spanish speakers, or to read Spanish text) are not able to operate when I am surrounded by monolingual English speakers and choose not to buy any books written in Spanish. That information, in the context of *my* particular mind and environment, has therefore very little effect on the world. Other representational content may not have much potential in any context: a poorly written novel, which neither stirs the heart nor stimulates the mind of the reader, will struggle to survive in the competition for our attention. There may be some mileage in being *associated* with a successful replicator (i.e., being selected as a side effect of a replicator with useful effects), but in general a meme demands content that has an executive role, in (potentially) producing a phenotypic effect.

From this it is apparent that there is a clear distinction between a replicator's *content* and its *effects* on the world: memes must be *about* the things that they affect, just as DNA can be said to carry information *about* the phenotypic effects that genes control. The key question for memetics, then, is this: in what form might units of cultural information be preserved, so that the content of those units can both be maintained between generations and at the same time be able to produce the relevant effects on the world? DNA does it for genes: what is the equivalent for memes?

Representational Content: The DNA of Culture

This chapter introduces the thesis that memes – the units of cultural information – should be specified by their *representational content.* What exactly does this mean? At one level the answer is straightforward: it simply means that, as *representations* of a portion of information, memes can be said to have a certain *content.* "Representation" is not a word that occurs frequently in the vocabulary of most people, but philosophers do not mean anything very complicated by it. Human minds are furnished with all sorts of mental states and events, including thoughts and feelings, attitudes and opinions, memories and skills; a "representation" is simply some piece of our mental furniture which carries information about the world. For example, a thought that "the object on my desk is a book" is a mental representation of a bit of the world (i.e. that book). So "representational content" refers to the information that is included in the content of our representations.

The complications arise when we start to ask how we know exactly what information is included in any given mental representation. In the example given, exactly which bits of information about the book are included in my representation of it: the fact that it can be read, that it is a paperback, that it is a dictionary, or what?

Philosophers' various responses to this problem are known as "theories of content". There is as yet no consensus on which of these theories is correct, but the resolution of this debate is of key importance to memetics. It is representational content, I shall argue, which accounts for the mechanisms of memetic heredity and for memes' power over their phenotypic effects, in the same way that the nature of DNA accounts for the mechanisms of genetic heredity and for genes' power over their phenotypic effects. Moreover, it is a meme's basis in representational content which enables it to carry information of the depth and complexity that we find in modern human culture, and to interact with the other memes in its environment. An adequate theory of representational content is as important for memes as an understanding of DNA is to genetics. Such a theory must be able to determine both which sorts of representations count as memes, and how we can specify the content of those that do.

Thus it is the task of the remainder of this chapter to extract, from philosophers' musings on this subject, the factors most relevant to memetics. In this sense it is a more philosophically "technical" chapter than the rest of the book, but the effort is worth it to clarify such an important issue. It is only fair that I should draw the attention of nonphilosophers to the controversial nature of some of what follows – in that the details pertain to a theory of content which would not be endorsed by every philosopher in this active area of debate. Nonetheless, the conclusions that I reach are not totally dependent on my preferred theory of content, which may be treated by those philosophers who disagree with it as merely illustrative of the fact that an adequate account *can* be given of how representational content ("memetic DNA") preserves information between generations in the appropriate way.

Representational Content – a Technical Interlude

When scientists were searching for the chemical basis of genetic heredity, their focus was on this question: given that our bodies do preserve information from generation to generation, how is that information physically realized, and how can we ascertain which information is contained in which bits of the physical structure? There is an analogous question for philosophers who want to know how the content of our mental representations is fixed: given that we humans do carry mental representations of the physical world (as well as abstract concepts, etc.), how do we obtain the information that they contain, and how can we ascertain which information is contained in which representations?

Simple Indicator Theory

I return, as a discussion example, to my representation of the book on my desk and the question how its content is determined. In other words, how do we know for certain which bits of information about that book are contained in my representation of it?

One answer is that the content of any belief is determined by the state of affairs that causes it, or that it reliably indicates. So the content of my "book" belief is determined by the book itself, with whatever properties it actually has – and indeed this accords well with common sense. Not many people would naturally fall to questioning themselves too deeply about the content of their mental representations, but many would share the intuition, once this issue *has* been raised, that the content of their beliefs must be determined by the bits of the world that trigger the beliefs in the first place. To put it simply, I *believe* that there is a book on my desk, because there *is* a book on my desk. In philosophy this view is known as the simple indicator theory of content.[1]

In a bit more detail, it goes as follows. Organisms have certain perceptual abilities. I, for example, can see the book on my desk. When I do so, something happens in my brain which indicates to me that this object has been perceived: a sort of mental "flag" is raised. More formally, this flag is called a natural internal indicator: it *indicates* that something has been perceived; it is *internal* because it's inside my brain; and it is *natural* in the sense that it happens purely as a result of my innate nervous system – I cannot choose whether or not to perceive the objects in front of me, and have no conscious control over what goes on in my brain when I do. This particular flag, then, is my natural internal indicator that a book has been seen.

Now, according to simple indicator theory, whenever I see a certain object – in this case a book – the *same* mental flag is raised in my brain; and this is a different flag from the one that is raised when I see Joan, or an apple. In this sense the "flags" carry information about the external situations that most reliably cause them to be raised. I know that what I have just seen is a book, and not a piece of fruit, because the flag that was raised in my brain is the one that reliably indicates the fact that I have just seen a book, and not the one that is raised whenever I see an apple.

Obviously it is very useful for me to have information about certain objects, and in lots of cases it will be equally useful for that information to prime me to behave in a certain way: greeting the object (Joan) rather than trying to eat it (the apple), for instance. So my mental flags are part of a causal chain, in which their being raised is the *effect* of relevant perceptual input, and is subsequently the *cause* of appropriate behaviours.

Thus the proponents of simple indicator theory conclude that the content of our beliefs is fixed in the following way: beliefs are those natural, internal indicators which have become representations with the function of controlling a certain behaviour, *because of* the information they carry about external situations, and *in order that* the behaviour may be produced whenever that situation occurs. To put this in terms of our example: the book flag is a representation whose function is to control various responses (the mental response "that's a book," the physical behaviour of picking it up to read it, etc.) *because of* the information that it carries about the external situation (i.e., because *this* flag gets raised whenever I see that type of object), and *in order that* the same responses may be produced whenever I see a book. So the information that is carried by this book-representation is determined by my past observations of books.

The Disjunctive Problem

Common sense, of course, is not always the best guide to reality, and the most notorious difficulty with such an account is known as the disjunctive problem.[2] The central claim of simple indicator theory is that the content of a belief is fixed during a learning period and determined by that which most reliably causes the belief. The disjunctive problem is that "there are many equally good ways of describing the conditions under which a particular representational state has been selected."[3] To put this another way, simple indicator theory says that the content of a belief is determined by whatever reliably causes it – but the difficulty lies in establishing *what* was the most reliable cause.

Problems arise, in particular, when a mental flag is raised by the perception of something that is – to the person or creature involved – indistinguishable from that which *usually* causes it to be raised. Perhaps the object on my desk is not a book in the conventional sense, but rather a box that has been designed to look very much like a book: when the "book" is opened, it reveals a hollow centre for storing valuables in a place where burglars would not usually think of looking. When I glance at this object, it looks so similar to a book that the "book" flag is raised in my brain. Do I now have a correct or incorrect representation of the object on my desk?

Common sense at first dictates that my representation is obviously incorrect. What I see on my desk is not a book, but a book-shaped box. But wait. Whether my representation is true or false will be determined by the match between the object that triggered it and the representation's content – which, according to simple indicator theory, is determined by whatever reliably causes it. Well, book-shaped boxes reliably cause this particular representation, just as ordinary books do. (Otherwise, such boxes would be pretty pointless and rather unmarketable.) In which case my representation is reliably caused, not by books per se, but by books-or-similar-looking-objects. This implies that what I have is a *correct* representation of a book-or-similar-looking-object, rather than an *incorrect* representation of a book.

Things look simpler if (as I believe to be the case) we can be sure that I have never in my life before seen one of these book-boxes, so that every previous time the relevant flag has been raised in my brain its cause has been an actual book. The most reliable cause of my representation is, then, an ordinary book, and it seems sensible to say that, this first time I come across a fake book, when it triggers the same flag my representation of it is just incorrect. Amongst other reasons, it is clear that this particular flag triggers behaviour that is quite inappropriate in response to a book-box (which cannot be read).

Yet what if I'm wrong in thinking that I've never seen one of these book-boxes before? What if in fact I *have* seen several of them, without realising it, and without even knowing that such an object exists? If this is the case, then I have a representation of certain objects in my environment, which is reliably triggered by both actual *and* fake books. So simple indicator theory would have to say that the representation's content is something like books-or-similar-looking-objects. Equally clearly, however, if I opened a book-box in the hope of reading it then I would be disappointed, and indeed would no doubt *say* that I had made a mistake (formed an incorrect representation).

It is at this point that many fine minds, unused to the sorts of thought experiment that fascinate philosophers, begin to find such discussions vertiginous. This particular example is brought to a standstill by the fact that of course you could *ask* me how I was representing the object on my desk; indeed, my very choice of words ("book" vs. "paperback" vs. "booklike object") would give you more than a small clue to the content of my representation.

The question behind the example, however, moves on. How is the content of a representation fixed, if not simply by that which most reliably causes it? In our search for the answer, we cannot yet forget about the disjunctive problem.

The Philosophers' Frog

The classical philosophical illustration of the disjunctive problem centres around a frog, who can perceive small black things (sbt's), but whose visual system is not sophisticated enough to be able to distinguish between different types of sbt: many of the sbt's in the vicinity happen to be flies, which are nutritious for the frog; but some (such as wind-blown grit, which the frog cannot distinguish from flies) are not.

The frog, then, has a natural indicator of sbt's. According to simple indicator theory, if flies are the most reliable causes of the indicator's being triggered, then that indicator carries information about the presence of flies. If, in addition, the frog is rewarded for flicking out its tongue whenever flies are present, then it will be useful for that indicator to be linked with the frog's tongue-flicking behaviour. Thus a representation will develop, with control over the relevant behaviour, and its content will be something approximating "fly".

Of course the problems arise because the indicator in question is not triggered *only* by flies, but also by every other sbt that the frog cannot distinguish from flies – even though some of these objects may not even be nutritious for the frog. So what happens on those occasions when the indicator *is* triggered by a different sbt? If all other sbt's in the vicinity are different but (from the frog's point of view) *indistinguishable* from flies, then how is it possible to tell whether what happens on these occasions is an *incorrect* representation of the sbt as a fly, or a *correct* representation of the sbt as a fly-or-other-sbt? To put it very crudely, when the frog flicks its tongue out in response to wind-blown grit, how can we tell whether what the frog "thinks" is wrong ("fly") or right ("fly-or-something-similar")?

Thus it begins to look as if the content of a representation cannot be determined quite as straightforwardly as simple indicator theory would claim. Situations like mine and the frog's show how hard it can be to characterize the most reliable cause of a representation, and this implies that there are many equally good ways of assigning content to it.

Ben's Lucky Mistake

Nonetheless, it is surely important not to become bogged down in arcane examples of the sort that hardly ever crop up in reality. Surely the vast majority of our mental representations are created as the result of encounters with a particular sort of object – a book, an apple, a person or whatever – and not as the result of encounters with a mixture of that object and objects indistinguishable from it.

This may be true, but unfortunately does not help very much, as my next example will show. It illustrates the case of representations that might be described as "lucky mistakes". In such cases, a flag is raised by something different (but indistinguishable) from what usually raises it, but – unlike when the frog catches some grit – the behaviour that it triggers is fortuitously appropriate.

Suppose that a wasp flies into a room where there is a small child who has never before encountered one. Ben has, however, seen bees many times before, and the wasp now triggers the representation that he has previously had whenever bees have flown into the room. That representation was created as a result *only* of previous encounters with bees, since this is the first time that Ben has seen a wasp. As a result of his beliefs about those sorts of insects, Ben will believe that this one might sting him, avoid antagonising it, and do his best to let it back into the garden. Unlike my response to the book-box (which would be inappropriate: any attempt to read it would fail), Ben's response to this insect is quite appropriate.

My intuition about Ben is that, rather than a correct representation (of a bee-type insect, for instance), he has made a lucky mistake (thinking it is actually a bee). As it happens, for the purposes of his behaviour, it did not matter whether it was a wasp or one of the things whose properties were actually responsible for his representation. In both cases he should want to act as he did. Indeed, he can quite happily go on through his life representing all such insects in the same way, for in functional terms this will lead him to successful behaviour in relation to them. In reality, however, Ben *has* made a mistake: wasps are biologically different from the other things with which he is co-representing them. If a more knowledgeable person had been with him at the time, then she might have pointed out the difference but advocated that he adopt the same strategy with both insects. It just does not seem relevant that for Ben's purposes at the time it did not matter whether the insect was a wasp or a bee, since he should have wanted to be rid of either one. No matter that both insects fulfilled the same purpose in his life (causing him to represent them in a certain way, and therefore to avoid them); what does seem relevant is that he assigned the wrong identity to the wasp. Wasps are not bees.

Such examples show that even in cases where it seems obvious how to describe the most reliable original cause of a belief ("bees" – because Ben had encountered no other such insects during the learning period), future events might lead us to question our descriptions (was it after all "bees", or was it "black and yellow buzzing insects", or was it something else?). Following hard on the heels of the frog's problems with sbt's, Ben's encounters with hymenopterans present a serious challenge to simple indicator theory. Does this mean that we have come no closer to an explanation of how the content of representations – even quite simple ones like Ben's and the frog's – are fixed?

Types of Property

Fortunately not. Rather, what we can learn from such examples is that we need to look in more detail at the objects that trigger our representations, to discover which of their *properties* are relevant. In other words, instead of thinking of a representation as containing information about particular objects or events, it will prove more useful to try to specify exactly which *aspects* of those objects or events are included in its content. All objects have a collection of properties – size, shape, colour, and so on – and only some of those properties will be responsible for triggering our representations of them. If we can identify these properties, then we can begin to specify the content of each representation and hence to discern which objects or events are (in)correctly represented, when they trigger it.

It will be helpful, in particular, to distinguish between two sorts of property: the functionally relevant properties of what is represented, and its causally relevant properties. Batesian mimicry neatly illustrates the difference between the two.

In Batesian mimicry, a harmless species is protected from predators by its resemblance to a harmful species. The venomous coral snake, for example, is mimicked by several other snake species, such as the harmless milk and king snakes. The coral snake has distinctive alternating yellow, red and black bands, and predators soon learn to avoid snakes with that appearance. Snakes which resemble the coral snake will benefit from this avoidance behaviour, even if they are themselves wholly harmless.

We can call the coral snake's venom a *functionally* relevant property of that snake. Its venom is the reason why, in the first place, predators learn to avoid the coral snake. Its appearance, on the other hand, we can term a *causally* relevant property. It is this which, in the future, will cause predators to avoid snakes which share that appearance.

Looking back at the wasp that flew into the room earlier, we can see that the *functionally* most relevant property of such insects is the knowledge that they sting: this is the reason why Ben's representation of them has gained control over his avoidance of them. In contrast, what is *causally* relevant to present or future encounters with them is their appearance: that is what now triggers the controlling representation.

This distinction is useful because it will help us, later in the chapter, to identify the properties that form part of any given representation's content. If, as I claim, representational content provides the mechanism for memetic evolution, then it is essential to be able to specify how that content is fixed in our representations (just as gene theory needs to know how genetic information is fixed in DNA). Replicators preserve and copy specific portions of information, and an adequate theory of the mechanism that enables them to do this should also tell us how to identify precisely which bits are carried in each replicator. In the case of memes, this means pinpointing the exact content of any given representation, and this will be determined partly by the various properties of the object or situation being represented; so the ability to pinpoint its relevant properties is crucial.

Yet this is not the only factor that will determine representational content. Content is fixed, in addition, by the capabilities and history of the organism doing the representing.

Different Sorts of Representation

It is reasonable to assume that not all representations will have the same level of complexity. We know from experience that our own representations can be constructed in more detail as time goes on. At first a familiar face is just that; then it might be associated with a name; and as we come to know that person better, extra layers of information are added – so that when she walks into the room today, the representation that she triggers is much more complex than it was last year. Equally, an expert in any field will have more complex representations of the concepts and entities within that area than the majority of lay people: contrast a civil engineer's mental representation of a road with that possessed by most of the rest of us.

So it is fair to say that I have some simple and some complex representations; and that of the representations that I share with Beth some of mine are simpler and others more complex than hers. From this we can see that representational complexity varies, not only between individual members of a species, but also within those individuals. How much more, then, must it differ from species to species, since different sorts of organisms are capable of such varying levels of comprehension of their surroundings (contrast an eagle's eyesight with a bat's hearing and a primate's social awareness).

If we are to discover not only how representational content is specified but also which sorts of mental representation might count as memes, then we need to gain an understanding of the different sorts of representation that can be formed – and to achieve this, we need to explore some of the *ways* in which they are formed.

Nonassociative Learning: Representations as "Switches"

Representations are formed, on the whole, to link a behaviour with the perception of a particular object or event. At the most basic level, such links are formed via a process known as nonassociative learning, in which learned behaviour is simply the result of exposure to a stimulus.

One example of this sort of learning is "imprinting" in young animals, who very quickly learn to recognize and be attracted to members of their own species, simply through exposure to their presence at an early enough age. Indeed they will become equally attached to a surrogate if it is presented at the right time: it is not unusual to come across stories in the press of orphaned chicks becoming attached to a family dog, and this is the process at work here. Habituation by much simpler organisms, like sea snails, is another instance of the same sort of process: these animals usually withdraw their respiratory organs when poked, but if exposed to repeated stimulation then they become habituated and stop withdrawing.[4]

The key in all such cases is that an inherited tendency is given direction by exposure to a stimulus. Returning to our frog, its reaction to sbt's seems also to be an (imaginary) example of nonassociative learning: it has evolved such that early exposure to whizzing sbt's sets up a link between them and the tongue's response.

Thus the frog, like the sea snail in the real world, exhibits under the relevant stimulus a type of behaviour that might be described as "on/off": either the sea snail withdraws or it does not; either the frog flicks out its tongue or it does not. There is, though, one key difference between the two examples. The sea snail's behaviour is controlled directly by the stimulus, but in the frog's case there is an intermediate step, whereby the behaviour is controlled by a representation that is triggered by the stimulus. Under these circumstances, the representation can helpfully be seen as playing the role of a switch that turns the behaviour on or off. Clearly such a representation has a very low level of complexity.

Internal Properties

Returning to Ben's behaviour when the wasp flew into the room, it may be that this, too, can be viewed as "on/off": perhaps it consists simply in his keeping away from the insect, so that if he sees such an insect then he avoids it, whereas if he sees a black, fly-like creature then he doesn't bother. If this is so, then it is fair to describe his representation, too, as just a switch that controls his behaviour.

Alternatively, his behaviour may be more complex: he may avoid the insect, but also think to himself "there's a bee" or just "there's one of them again". In this case his representation is clearly more complex than our imaginary frog's. Not only are certain of the *insect's* properties (appearance, sting, etc.) relevant to Ben's representation, but so is something that is going on inside *Ben's* mind. His avoidance behaviour is triggered by an *external* stimulus, but his thought is about its identity, and this is an *internal* property. That is, the thought ("there's one of them again") relates this present perception to previous encounters with similar insects: it links one internal representation with other, earlier ones. Ben's representation is thus involved in a relatively complex system, in which other representations may affect the role that it plays in controlling behaviour.

Internal Properties and Lucky Mistakes

It is internal properties, such as identity, which provide the key to understanding cases of "lucky mistakes". Whereas some representations simply play the role of a switch in turning a given behaviour on or off, the content of others includes not only the external properties of that which stimulates them, but also internal properties such as identity. Consequently, if we want to discover whether a representation has been triggered correctly in such cases, then we need to look not only at the stimulus and resultant behaviour but also at the representation *itself*.

If the role of a representation in controlling behaviour is simply that of a switch, then it seems right to say that the representation has been "tripped" correctly if it results in appropriate behaviour and not otherwise. This is because the only thing that we need to take into consideration in answering this question is whether the correct stimulus-behaviour link has been made.*

If, on the other hand, the representation's role is more sophisticated, then we cannot rely on the appropriateness of the resultant behaviour to reveal whether the representation was also appropriate. Going back to Ben and his wasp, for instance, it may be that the resultant behaviour (avoidance) is fortuitously appropriate, but that the representation itself is incorrect (because he has assigned the wrong identity to the insect). This is because its accuracy is determined not only by the appropriateness of the link between stimulus and behaviour but also by certain internal properties. There are links not only between this representation, an external stimulus and a given behaviour, but also between this representation and *other* representations – and to determine its accuracy, we need to check the appropriateness of *all* of its links, internal as well as external.

It is by now clear that the content of more complex representations is determined not only by the properties of the external objects that they represent (and we have broken these down into "functional" and "causal"

* Obviously things are not quite as straightforward as this implies: the disjunctive problem reveals complications in determining whether the representation was tripped by an appropriate stimulus, but the point is that the only links to be checked are between the stimulus, the representation and the resultant behaviour.

properties) but also by any internal links that the organism has formed between those representations and others. Since most human representations are obviously of this more complex sort, these conclusions will prove vital to establishing the nature of the representational content that forms the basis of memes.

Indeterminate Content?

Armed with the distinctions between "switches" and more complex representations, and between different sorts of property (causally vs. functionally relevant), it should now be possible to begin to track down the content of any given representation.

It makes sense to begin with the simplest sort, such as the frog's, which are formed by an innate response to stimulus exposure. Let's take just one of the flies' properties, and investigate how we can tell whether or not it is included in the frog's representation of the ambient sbt's. The property of being food seems to be a good candidate for investigation, since this is the key reason why the frog's tongue-flicking behaviour has become linked to its representation of that sort of object. In this sense, then, food is a *functionally relevant* property of the flies, and this would seem to imply that it must be included in the content of the frog's representation.

Again, however, things are not so straightforward as they might at first appear. Suppose that, once the representation is set up, the proportion of nutritious sbt's in the area changes (say because pellet shooting begins to take place in the area, and the frog cannot distinguish whizzing pellets from whizzing flies): whereas the majority of sbt's used to be flies, now most of the objects that the frog catches are indigestible. The frog's problem is that it still cannot distinguish visually between the two, and persists in catching all sbt's: its representation, which was formed originally as a result of exposure to flies, is now triggered largely by pellets. How, then, can we say with certainty which properties are included in that representation's content? Although food *was* a functionally relevant property of the objects that it initially indicated, *now* the representation is triggered mostly by objects that are not food items.

Technically, what the frog cannot do is modify its reaction to things with the same *causally* relevant properties, in response to a change in their *functionally* relevant properties. In other words, even when almost no sbt's are frog food, still it continues to catch everything with the relevant appearance. The frog's representation was set up to link its tongue-flicking behaviour with its perception of sbt's *because* of their nutritional value at the time. The problem is that the representation is now fixed in its brain: it does not have the learning ability to alter it in response to a change in the stimuli.

This means that it is impossible for us to find out whether or not "food" is part of the frog's representation. One could argue that it *must* be, because flies' nutritional value is the very reason why the representation exists in the first place. On this view, the frog is making a mistake whenever it catches a pellet. Equally plausibly, one could respond that it *cannot* be, because the frog behaves in exactly the same way, whether or not most of the ambient sbt's are nutritious. On this view its representation was set up to indicate things with the relevant appearance, and it does so correctly whenever a pellet whizzes past; it's just that the frog doesn't benefit in those cases. It is not possible to resolve the debate between these two points of view, however, because there would be no detectable difference in the frog's behaviour, no matter which were true. In a controlled experiment, in which sbt's of varying nutritional value were whizzed past and the frog's reactions monitored, it would respond to all of them in the same way. According to one viewpoint it would be making a mistake every time it responded to a pellet, and according to the other it would not – but observation would not tell us which was the case.

Of course it may be possible to test whether any of the stimulus's *causally* relevant properties is included in the representation: by whizzing different sizes, shapes and colours of object past the frog at different speeds, for example, we could observe which sorts of object elicit its tongue-flicking response, and in this way compile a partial list of the properties that comprise its representation. On the other hand, this still leaves open the question which of the *functionally* relevant properties should be included in that list, and to this extent a complete description of the frog's representation will remain elusive.

One solution may be to say that any functionally relevant properties are *potentially* elements of the representation's content. The fly's "food" property, for example, is potentially an element of the representation's content, because it was functionally relevant to the formation of that representation – but I would suggest that we need to ask a further question to determine whether it is *actually* included in that content. The further question is this: is the representing organism able, once the representation has been established, to modify its behaviour in response to a detectable variation in that property? Unless the organism is capable of such behavioural flexibility, then it is just not possible to list with certainty the contents of its representations. The frog continues to catch every sbt, even when circumstances change to the extent that hardly any sbt's provide it with food, and thus it is impossible to test whether or not "food" forms part of the frog's representation.

Far from tracking down the content of the frog's representation, therefore, the considerations in this section have revealed that simple representations of this sort may well have content which is to a certain extent indeterminate. What else, then, will be needed in order to determine with precision the content of a representation?

Associative Learning

In order to change its behaviour in response to such a stimulus change, the frog would need to be able to engage in *associative* learning. This differs from nonassociative learning in that it depends on the experience of an association or relationship between events rather than simply on stimulus exposure. A classic example is Pavlovian conditioning, typified by the experiments in which Pavlov's dogs would drool at the ringing of a bell because they had learnt to associate its sound with impending food. The sound of a bell ringing had no intrinsic properties that would benefit the dogs (unlike the flies whizzing past our frog), but they learnt to associate its sound with something that did benefit them: food. What they developed, then, was a response to an association between events (ringing bell and the arrival of food), rather than simply to an event itself (the arrival of food).

Learning of this sort removes the problem of indeterminate representational content for the following reason. Representations, in such cases, have been formed *because of* the animal's awareness of the link between stimulus and reward, which means that if there is variation in that reward then the animal can modify its response accordingly. If food stopped arriving whenever a bell was rung, for instance, then the dogs would lose their salivating response to the bell. To put this more formally: if there is no change in the causally relevant properties of the stimulus (it maintains its sound, appearance or whatever), but nonetheless there is a detectable difference in the functionally relevant property (i.e., the reward), then that change would break the creature's association between stimulus and reward, leading to an observable behavioural change.

The difference between this and the more "switch"-like situation of the frog, is that here the external links (between stimulus, representation and behaviour) are not the only sort. The dog could not engage in associative learning unless it were aware of the link between bell and food – or in other words if there were not also *internal* links between its representations. It is these internal links which will enable us to establish whether any given property of the stimulus forms part of the dog's representation – for we can change the association between the stimulus and the property that is under scrutiny, and observe what effect (if any) this has on behaviour. If the animal's response alters, then clearly that property must form part of its representation. If there is no behavioural change (in a creature which is capable of making such a change), then we can say that the property is irrelevant to its representation. The key point is that, so long as an organism can modify its behaviour in response to variation in a given property, we can determine whether that property forms part of its representation.

The Story So Far

Representational content has been hypothesized as the cultural DNA – yet if there are various levels of representation, not all of which are even determinate in terms of their content, then what can representational content tell us about memes?

Darwin's theory of natural selection needed a mechanism via which information about physical characteristics could be copied from one generation to the next. Mendel's genes provided the answer, and with the discovery of their basis in DNA scientists were able to account for the preservation and replication of that information, its control over phenotypic effects and its capacity to exert that control in a variety of contexts. The theory of cultural evolution, too, needs a mechanism via which cultural information can be preserved in a way that enables it to be replicated and to exert control over its effects in a variety of contexts.

This chapter has shown that there are different sorts of mental representation, with some being more complex than others. In particular, some representations play a role much like a switch, linking an organism's perception of a given stimulus to behaviour that is appropriate as a response. Others are more complex, and have not only these external links to perceptions and behaviour but also internal links to other representations – and the content of any given representation will be determined by all of these links.

This means that it may be impossible to specify completely the content of one of the simpler sorts of representation. On the one hand we can alter the properties that we surmise are *causally* relevant to that representation, and deduce from our observations of any resultant behavioural changes whether or not those properties form part of the representation's content (perhaps the frog does not respond to ambient whizzing black things over a certain size, for instance). On the other hand, if the organism is incapable of responding to changes in the stimuli's *functionally* relevant properties because it is limited to nonassociative learning methods, then there will be no behavioural changes to observe. A complete list of properties included in the representation's content will therefore remain elusive.

With regard to the more complex representations, however, we should in principle be able to complete the list. The internal links between the organism's representations will enable it to partake in associative learning: to respond not only to new stimuli but also to the associations that it makes between those stimuli and other events or entities. It will thus be possible to determine which properties make up the representation, by changing both the stimulus (e.g., bell) and its associations (e.g., food): if the organism is able to respond to these changes, then we can begin to track down the content of its representations.

Representational Content: The DNA of Culture

Thus we can see that the content of a representation includes those causally and functionally relevant properties to which an organism can adapt its responses. Representational content, on this view, is determined by an interaction between the relevant properties of that which is represented and the learning capacities of the organism involved. Some organisms are able to represent the world around them, even though they are essentially stupid and preprogrammed, and even though it may not be possible to specify exactly what the representation's content is in each particular case. Their representations would obviously not count as memes.

There are two reasons for this. First, if the content of their representations (or in other words the information that they carry) is indeterminate, then those representations do not have one of the most crucial aspects of any unit of selection: the capacity to keep information intact from generation to generation.

Secondly, even if they did have determinacy of content, then this in itself would not be enough to establish their status as memes. Evolution demands not only the preservation of information but also its replication. Even if an organism had the sort of behavioural flexibility and learning capacity that enabled it to develop representations with a fully determinate content, still there would be a further question to ask: is the organism able to modify its behaviour in response not only to changes in the stimulus and its associations, but also to variations in *other* organisms' behaviour? If not, then the sophistication of its representations will count for very little: there is no replicative mechanism available to them, and therefore they are not memes. Their content, depending only on that particular organism's nervous system, will have no collective causal history.

This three-stage process (Is there any content? Is it determinate? Is it replicable?) for determining the content and status of a representation has turned to some extent on the differences between individual learning methods. Additionally, when discussing the mechanisms of memetic replication, not only individual but also social learning methods will be at issue. To be capable of copying and retaining memetic information, an organism needs the capacity to adapt its behaviour and its representations in response to observations of others' reactions – to be able, in other words, to engage in certain types of social learning. The question which types of social learning are adequate to the task of memetic replication is discussed later, in Chapter 9.

So: some organisms are capable of forming representations whose content is determined by a combination of the relevant properties of that which is represented, and the organism's own individual and social learning capacities. Such organisms are able, in other words, both to preserve information and to transmit it between themselves. What else is needed for the content of their representations to play the role of cultural DNA? Information is useless unless it can be implemented: cultural heredity depends on memes' ability to preserve and copy information *which can then be put into effect.* Representational content must, therefore, be able to account for the actions and reactions that stem from it, if we are to believe that it constitutes cultural DNA.

Fortunately, this is exactly what representational content does best. Representations are formed when behaviour is modified in response to a stimulus: in other words, representations are specifically those bits of our mental furniture which control behaviour as a response to incoming information – exactly what meme theory demands. The *content* of those representations is determined by the properties of the stimulus which are causally or functionally relevant, and to which the organism is able to modify its behaviour. The relevant behaviour then occurs *because* of the content of the representation that controls it – or in other words it is the representation's content that determines which effect it controls.

In particular, when it comes to the sorts of complex representations that might count as memes, their content depends on their playing a certain *type* of role in controlling behaviour: they must be able to affect and be coordinated by other representations – otherwise internal properties (e.g., identity) lose their significance. This explains how representational content is able to account for memes' general applicability: a representation is bound to have a wide range of applications, since the way that its content is determined *depends* upon its ability to interact with other representations, and be modified in response to changes in them. In other words, a meme will be generally applicable *because of* the nature of its content.

Conclusions

This chapter began with a challenge to memetics: how does cultural heredity work? One possible answer to this question – and I make no claims to be culture's Watson or Crick, but simply to have demonstrated that it *is* possible to give an answer – is that representational content is the cultural equivalent of DNA. It is important to be able to specify what memes are, because a crucial element of evolutionary theory is replicators' independence from the effects that they control. The job of replicators is to preserve information in a way that enables it both to be replicated and to produce its effects – and this chapter has shown how representational content allows memes to fulfil that role.

This is despite the fact that not all representations will count as memes. Some are more complex than others, and in the case of the simpler ones it may not even be possible to give a complete account of their content. If a representation merely plays the role of a "switch" that turns certain behaviour on/off at the perception of a given stimulus, then the only ways of checking the content of that representation will be to examine the behavioural results of varying the stimulus. If an organism cannot respond to changes in the functionally relevant properties of the stimulus, then it will not be possible to ascertain whether those properties are part of the representation.

In the case of more complex representations, however, which have links not only externally to perceptions and behaviour but also internally to other representations, the resultant behavioural flexibility will enable us to track down their content more completely. It will be possible to test all of the links, by altering the associations that the organism encounters and observing the effects on its behaviour.

Only representations with this determinacy of content can count as memes, since a crucial aspect of any replicator is the preservation of given information. Such determinacy will depend as much on the organism's faculty for individual learning – to form links between its various representations – as on the properties of that which is represented. These internal links will, moreover, prove crucial to our understanding of the difference between the simple cultural inheritance that might be found in other species, or in our ancestors, and the fully fledged memetic evolution that is found only in modern human culture.

In addition, representations must be replicable if they are to count as memes, and this too will depend partly on the organism's abilities – in this case on its capacity for social learning (explored in later chapters). Finally, memes must be able to exert their effects if they wish to be selected, and the nature of representational content, as explored in this chapter, will enable them to do just that.

This chapter has strayed quite deeply into the territory of philosophical analysis, whose sometimes arduous terrain has nonetheless held, I hope, some allure. Things become a little easier from now on, because we have reached a point from where the rest of meme theory can advance without any further need for the apparatus of quite such technical discussions. It was important, though, to begin with an account of how memetic information can be preserved between generations, for without this the meme hypothesis is baseless.

4

The Replication of Complex Culture

In order to replicate, memes need to be able to *pass on* as well as to preserve their content. The key question here is not so much *which* copying mechanisms support the spread of memetic information, as how *any* such mechanism can support the immense complexity of human culture. Any account of cultural development must include an explanation of what has enabled this complexity to increase and persist. If memes are the units of cultural evolution, then their replication methods must be able to sustain the enormous breadth and depth of information that has built up over the millennia, and meme theory must be able to account for how this happens. Following a brief look at the ways in which cultural information spreads, the bulk of this chapter is therefore given over to an examination of the key features of the replication of complexity, investigating how it might work in principle as well as how it is played out in practice, in culture as in nature.

How Is Cultural Information Copied?

Imitation seems to be one of the most obvious methods by which cultural information spreads: I might learn a skill from one person by observing her actions, or pick up the musical style of another by listening to his recitals. In addition, however, there is often an intentional element in our learning. We are constantly engaged in a process of deliberate communication with each other, and this is surely the most frequent method of cultural replication. I can gain new ideas and skills from you in ordinary conversation, and it will be apparent that I have acquired some novel information when it has an effect on my behaviour, opinions or future conversation.

An even more deliberate form of copying information is, of course, formal teaching, and teachers employ a variety of teaching methods. It is fashionable to bemoan the loss of the skill of rote learning – and a sound grasp of any subject is of course dependent on the acquisition of a certain amount of basic knowledge – but the best teaching does not impart information for pupils to regurgitate parrot fashion, but rather transmits an understanding of methods or principles. The aim of a mathematics lesson, for example, is not to ensure that pupils memorize the solutions of particular problems but rather to teach them how to solve any *new* problems that they encounter. In order to achieve this, they must acquire a generally applicable concept or method; the particular solutions are then the *effects* of having acquired that general method.

In other words, the content of what is learnt is generally applicable and has executive effect in producing particular solutions, artefacts and so on. All of this is consistent with the claim that teaching and other means of communicating are forms of meme transmission. Moreover, as with the analogous process of gene transmission, the copies will not always be exact, and the idea or skill in question may change in some way en route. The results of such a mutation might be the alteration of the words of a popular song, a gradual change in the way that dry stone walls are built, or the potentially disastrous alteration of a chemical formula. The essence of the process, however, is that the pupil or listener acquires a representation of the relevant information.

I should mention that the content of this section is a lot more controversial than it might at first appear. Imitation, for instance, is not so much an obvious, basic process of meme transmission as a source of great controversy within memetics. Some commentators (most notably, Susan Blackmore)[1] have in recent years claimed that imitation is the *only* form of learning via which meme acquisition can truly take place, and such claims have given rise to much heated debate, within memetics, about the nature of imitation and its capacity for meme transmission. I shall delve into this debate, which is clearly of some significance, in Chapter 9. For now I want merely to acknowledge that there *is* a debate to be had – that whilst it is obvious that culture spreads from person to person, it is far from obvious either how that happens, or whether its methods count as memetic.

The Replication of Complexity

In a way, though, the debate about how meme transmission might operate in practice is of secondary importance for memetics. The key question is rather what are the essential features of any successful method of replication – the features that will be apparent in every instance, whether it be from nature or culture.

One of the most astonishing aspects of both realms is the enormous complexity that has developed over time. Any convincing theory of cultural evolution must therefore be able to explain – in the way that genetics does for the natural world – how such complexity can be replicated in a way that preserves its content from generation to generation.

It is well known that complex replication will always be more successful if the complexity involved is hierarchical. This fact has been neatly illustrated by Herbert Simon,[2] in a parable which suggests "a general functional reason why complex organization of any kind, biological or artificial, tends to be organized in nested hierarchies of repeated subunits".[3] It goes roughly as follows.

Each of two watchmakers has to assemble watches from a thousand component parts. He has, in effect, to replicate an established complex form. Tempus assembles his watches piece by piece, and they are so constructed that if he pauses or drops an unfinished watch then he has to start again from scratch. Hora, on the other hand, makes subassemblies of ten parts each, then subassemblies of ten of these, and finally a whole watch from ten of those, so if he is interrupted then he loses only a small part of his work. As a result, Hora can assemble his watches in a fraction of the time that it takes Tempus: according to Simon's analysis, if there is a chance of say one in a hundred that either watchmaker will be interrupted while adding a part to his assembly, then Tempus can be expected to take four thousand times as long as Hora to assemble a watch. Although in fact the statistics of expectation show that the correct relationship is more like two thousand times as long,[4] Simon's key point still holds: the Hora style of building gives a better time scale, greater stability and resistance to shock, and a greater amenability to repair and improvement – and it is clearly hierarchical.

Dawkins not only agrees with, but has gone on to develop Simon's hierarchical argument,[5] "concluding that the evolution of statistically 'improbable assemblies proceeds more rapidly if there is a succession of intermediate stable sub-assemblies. Since the argument can be applied to each sub-assembly, it follows that highly complex systems which exist in the world are likely to have a hierarchical architecture'."[6]

The significance of assemblies* for the natural world is that the form of an organism must be structurally stable and, among complex forms, assemblies are the ones that have had time to develop. In particular, assemblies are needed to organize the *replication* of successful characteristics between generations. If characteristics were allowed to vary within subsystems (such as temperature control, digestion, etc.), then any amount of unsatisfactory combinations might occur. If, on the other hand, whole subsystems are replicated, then structurally stable complex forms will be able to develop. There are patterns of dependence within an assembly even though the individual constituent units may have been of no use if they were replicated independently.

Replication, in other words, will be most efficient if it builds on what already exists rather than starting afresh each time. An important implication of this message is that the most successful sort of replication will be particulate: if the constituent parts of what is replicated were to blend, then the end product would be a conglomerate rather than an assembly. The units of an assembly must be what Arthur Koestler[7] has described as "self-assertive": each maintains its own individuality within the assembly. On the other hand each must also be compatible with the others in the assembly, otherwise the result would be unstable: as part of a larger system, towards whose future and stability they tend to "work", the units in an assembly must (in Koestler's terms again) be "integrative" as well as self-assertive. The replication of complexity, in its reliance on assemblies, is therefore dependent on the existence of dual-natured units, which are able to retain their individual identities whilst operating as part of a complex.

In order to achieve such a complex, Koestler also highlights the fact that the units must be governed by certain functional and constructive rules. The structure, stability and behaviour of an assembly can be understood as the result of a set of invariant rules, although variation will be allowed in the "strategies" that are actually employed – just as in chess, for example, there are invariant rules that govern how each piece may move, but the actual moves or strategies employed during a particular game will be variable. So the assembly's rules govern which function or structure *can* be followed, and the strategy determines which *will* be followed.

* Dawkins's term comes with less baggage than the alternative, "hierarchy".

These two features – the self-assertive/integrative tendencies of units within assemblies, and the rules/strategy compromise of their overall organisation – enable the content of complex assemblies to be replicated with relative fidelity between generations.

Genetic Replication

This rather abstract account of the replication of complexity can be observed, in practice, in the behaviour of genes. It is clear, for example, that genes have the requisite dual nature. Dawkins called them "selfish" because they have the effect of promoting their own welfare at the expense of other genes in the pool and, more generally, they show self-assertive tendencies by the simple fact that they retain their particulateness throughout replication. On the other hand, genes have a better chance of surviving if they form coadapted complexes, and in particular they have tended to "band together" in survival machines to maximize future preservation: a clear display of integrative tendencies.

We may also look at the rules which govern the replication that gives rise to their assemblies. A gene's function and structure are fixed, but exactly how it does function will depend upon its environment. So, for example, the invariant rule of a gene for blue eyes is that the only effect it *can* have is on iris colour; but (being recessive) it *will* only have that effect in practice if it is passed on to an individual in which there is no "brown eyes" allele. Although governed by a strict rule, then, a gene's actual behaviour may vary.

It is not only the genes' phenotypic effects that are governed in this way: their interactions with each other also vary within strict boundaries. The creation of a sperm or ovum is governed inflexibly by the rule of "crossing over": parts of each paternal chromosome interchange with exactly corresponding parts of each maternal chromosome. Nonetheless, what actually happens during each instance of crossing over is so flexible within this rule that each created sperm or ovum is practically unique. Although the genes involved in this creative process are acting in accordance with an inflexible canon of behaviour, the end result is different on every occasion.

Memetic Replication

If genetic transmission is subject to assembling constraints – and having argued that such constraints will be true of *any* successful method of complex replication – it is now time to ask whether memes do not also operate within organized assemblies. If not, then the onus is on meme theory to explain how culture has managed to evolve such complexity, so swiftly. Notice that there is no suggestion that assemblies are an essential feature of *any* sort of replication: rather, where there is evidence of complexity within an evolutionary system, we should expect that complexity to be the result of assembling constraints.

Consider, first, whether memes might have a dual nature of the sort described. With regard to their self-assertion, it is clear that memes must, if they are truly units of cultural selection, share genes' resistance to blending. Unfortunately for memes, this is as yet one of their more controversial aspects – but later in the book I shall defend the view that they are, indeed, particulate, arguing in chapter 11 that memes may be distinguished from each other in the same way as Mendel originally distinguished genes.

Memes are also integrative, however. I have already noted that a new meme will have a greater chance of penetrating the existing meme pool if it is consistent with the others in that environment. For instance, the "earth is flat" meme is, in a modern context, much less fecund and long lived than in some previous centuries. In order to survive, memes – like genes – form coadapted complexes which display their integrative tendencies.

It is of course important to acknowledge that there is a difference between the acquisition of information and the acceptance of that information into one's network of beliefs. There is a sense in which the "flat earth" meme is still pretty successful today: plenty of people know that it is possible to believe that the earth is flat, even though they themselves do not subscribe to that belief. In this way it is perhaps analogous to a recessive gene, whose DNA we possess and are able to pass on to our children but which exerts no effect on our bodies or behaviour. Similarly, we are capable of passing on information that persists in our memories, even when we don't assent to it, but it will have little or no effect on our thoughts or behaviour. Just because I understand what it means to believe that the earth is flat, I neither subscribe to a conspiracy theory about the origins of satellite pictures of the earth nor have any doubt that it is possible to circumnavigate the globe.

Nonetheless, it remains the case that memes are integrative in the sense that they do best when they fit in with the others around them. There are two reasons for this. First, novel ideas that accord with accepted theories are more likely to be *remembered* than those which do not: I happen to have retained the flat earth meme, but there are plenty of other bits of information that I have forgotten because I deemed them invalid, or have lost because they are not called for in my present environment. Secondly, information is more likely to be *replicated* if it is absorbed into a network of accepted ideas or is useful in the context of much-used skills: we tend to pass things on more when we approve of them than when we do not. Unlike genes, which we receive as a job lot from our parents – so it does not matter whether they are recessive, and unable to exert their effects in some contexts – memes are copied on an ongoing basis throughout our lives. Being recessive is therefore more damaging for their prospects than it is for genes'.

So memes may be described as both self-assertive and integrative. Are they also governed by fixed rules – regarding their phenotypic effects, for instance – within which particular variations may be observed?

Consider as an example the "ability to read piano music" meme. Its phenotypic effect will, indeed, vary according to the environment (whether there is a keyboard and/or sheet music present) and to the rest of that person's meme pool (whether he also has the "ability to play the piano" meme). Nonetheless that meme will, given the suitable physical and memetic environment, have a fixed phenotypic effect: its possessor will be able to translate the written music into its physical expression on a piano. So it is also governed by the requisite fixed rule.

In such ways assembling constraints may explain the dynamics of meme transmission. Simon's parable demonstrated that complex replication via assemblies will be the most successful, and this should apply as much to memetic as to genetic units of replication. The discussion of cultural replication has remained a little abstract so far, but we can see the process at work in practice, in an area like science. If scientists rejected the whole of a thesis every time a contradictory result occurred, then no substantial theories of the natural world would ever have time to develop. Instead, scientists build on their existing knowledge: they replicate whole subtheories as parts of new hypotheses. They add to their present knowledge, collate it, discover its consequences, and in time a more complex thesis emerges. Sometimes (cf. Kuhn's revolutions) theories become extinct and the evolutionary process has to begin again. Memes compete continuously for the scientists' attention, and if they are to survive then they must fit into the functionally organized assembly of existing theories. Hence this assembling replication goes a long way towards explaining the dynamics of meme reception and retention. Memes will be acquired and retained if they are compatible with the existing assembly of knowledge (representational content) of the recipient.

It seems almost trivial to assert that we build on our existing knowledge, that it would be much harder to learn something entirely from scratch (contrast the effort needed by a complete novice in order to learn to play the mandolin with the relative ease with which a violin player would acquire that skill), and that what we already know affects both how and what we newly learn. Nonetheless, such mundane observations serve to support the hypothesis that memetic replication is a process of assembly.

There will, of course, be differences between the methods of gene transmission and those used by memes. In particular, genes are forced to replicate the whole assembly at once, but we are able gradually to acquire the contents of our mental assembly. Both sorts of transmission are, however, constrained by rules: a meme must, in order to be acquired, be able to slot into the established assembly just as much as any gene must, in order to be transmitted, fit into the assembly which is being replicated as a whole. With memes the process resembles the construction of a jigsaw, whereas with genes it is more like taking a photograph, but the essence of both procedures is the same: both sorts of replicator must maintain their particulateness whilst submitting to the assembly's rules of behaviour and construction.

Memes Versus Genes

The previous sections have shown how both genes and memes engage in the most successful methods of complex replication: processes of assembly that have their basis in dual-natured, variant-rule-following units. Since complexity is undoubtedly to be found within minds and cultures, we are justified in arguing that the essence of memetic replication must be a process of assembling representational content. Via teaching, communication, imitation and other forms of social learning such as gossip and normfollowing, the complexity of the cultural realm has been constructed over time, and the advancing complexity has been accelerated by the assembling constraints that govern the replication involved.

If the assembling constraints have accelerated the development of complexity for memes, however, then this raises questions about the difference in speed between evolution in the biological and cultural realms: if both depend on a similar process of replicating complexity, then why should memes have evolved so much faster than genes?

One answer is that the difference in time span between the two is less to do with the principles of replication involved than with the physical processes of each. Both rely on the assembling processes on which the replication of *any* type of complexity depends, so the principles involved are the same in each case. The underlying processes, however, are crucially different from each other: simply, a gene depends on its possessor reaching sexual maturity in order to be replicated, whereas a meme can be replicated again almost as soon as it has been copied. Since its replication rate is so much faster than that of genes, its evolution can be correspondingly faster, too.

Yet it is interesting to note that this difference between the two types of replicator does not apply across the board: "viruses and bacteria reproduce themselves much more rapidly than the vast majority of memes,"[8] and there are of course memes that are copied very slowly, too – such as traditions that are limited to a particular family and thus copied only once per generation, at the same rate as that family's genes. It is also worth pointing out the distinction between rates of *reproduction* and rates of *evolution*, since it is not inevitable that these should be linked: if errors in the copying process are rare, then even the speediest rates of replication will not give rise to great evolutionary change; if errors are frequent, then even slow reproduction can lead to relatively fast rates of change.

On the whole, though, it is clear that our memes are able to replicate much more swiftly than our genes, and that this goes some way towards explaining the furious speed at which culture has developed, relative to the natural world.

Conclusions

Memetic replication, then, must be dependent on the same sorts of assembling constraints as those involved in genetic replication, for it has been shown (by Simon, Dawkins and Koestler, amongst others) that such constraints are the most successful methods by which complexity can be replicated. This structured way of copying units of information (preserved in representational content) may therefore be seen to underlie the everyday cultural processes of communication, teaching and imitation. An explanation is thus provided for both the speed of cultural transmission and the expansion (rather than replacement) of its content over time.

This chapter and the previous one have revealed how memetics can account for the preservation and replication of complex cultural information. Yet the content of culture is not static but constantly changing: in order to evolve, there must be variation as well as replication in culture.

5

Variation

Few copying processes are accurate enough to rule out the possibility of error. This chapter addresses the questions how and why cultural information varies as it spreads, and whether there are any limitations on the changes that can occur. In particular, it focuses on two of the possible causes of variation – mutation and recombination – and on the vexed issue of memetic alleles.

Innovation and Genes

For genes, variation occurs in two ways: mutation and recombination. Mutations are sudden changes in organisms' characteristics, resulting typically from alterations in the structure of genes or chromosomes, which have the potential to be passed on to offspring. Recombination occurs when genetic material of different origins is mixed together: you have two alleles of each gene, and your offspring will receive only one of them, together with one from their other parent; they may therefore exhibit traits not seen in either of you (e.g., two brown-eyed parents might produce a blue-eyed child).

Nevertheless, there are limits on the innovation that genes can produce by either method. Why is this? There are obvious limitations on possible recombinations, in that alleles have to correspond and the possible recombinations are, though rich, finite with respect to any given gene pool. Popularly, however, genetic mutation is referred to as "random", with the unspoken implication that just about anything is possible. In fact, unless used with care, this term may easily lead to confusion. The important sense in which genetic mutation *is* random is that it entails "no general bias towards bodily improvement":[1] evolution's tendency towards improvement stems solely from natural selection. Nonetheless, this lack of bias does not mean that all changes are equally likely, and there are several senses in which genetic mutation is *not* random.

It is simply untrue to say that any change is equally likely, if by this it is meant that mutation will produce any event that selection might possibly favour. There is bias in favour of some changes and against others. In the first place, genes have varying mutation rates: some are more likely to miscopy than are others, and these rates of change may also be increased by external mutagens. There is also a bias involved in the *direction* of mutation, with some directions being more likely than others. Then, even once genetic mutations do occur, their consequences on bodies are restricted by existing embryology (recall that replication is subject to the restrictions of assembling constraints). Dawkins asks why, for example, birds' wings have developed in the way that they have, rather than in the style popularly attributed to angels. His reply is that, even though it would have been useful for birds to maintain a free set of forelimbs, "there may not be anything in the embryology of backs that lends itself to 'sprouting' angel wings. Genes can mutate till they are blue in the face, but no mammal will ever sprout wings like an angel unless mammalian embryological processes are susceptible to this kind of change."[2]

Such details apply to genes in particular. More generally, it should be noted that "random" is a context-dependent term, used to indicate that its subject is free from a *specific* form of causal control. If it is used independently of that context, without specifying from which kind of control a process is free, then the term becomes confusing and perhaps even meaningless. A process may be random with respect to one form of control but not to another. For instance, if you were asked to choose ten people "at random" from the street where I live, then it so happens that your sample could be fairly random with respect to age, sex and income – but could not be at all random with respect to colour (since all the street's residents are white). Or, if I asked you to pick a card "at random" from a pack that is scattered face down on a table, then I would really mean that it should be random with respect to its value and suit: it would not matter if you had a bias towards cards near the edge of the table, or towards cards that were partially hidden under others.

The best sense to make of "random", as applied to genetic mutation, is that it is not intrinsically biased towards increased fitness (on the contrary, most mutation is probably fatal). This, though, is not equivalent to the claim that all changes are equally likely, which would imply that genetic mutations somehow fall outside the realm of induction. In fact, the mutations that can occur are limited by the nature of what already exists – most significantly, by genes' mutation rates and by embryology. In addition, genetic recombination is constrained by the fact that genes may only cross over with corresponding alleles. Finally, the variations available to selection are the ones that are permitted by previously evolved developmental processes.

Innovation and Memes: Mutation

Returning now to the subject of memes – and although it would obviously be inappropriate to try to extrapolate from the detailed structural ways in which genetic variation occurs to the ways in which memetic variation might occur – it does seem that mutation and recombination are also widely observable in culture. In the absence of any other immediate candidates for the same role, their study will therefore provide a useful springboard for the investigation of memetic variation. I begin with mutation.

It has become clear that one of the most important elements of mutation is its randomness with respect to fitness – but that the restrictions on that randomness are also important. In this chapter I consider examples which demonstrate that such restrictions also apply to the elements of our culture which might be characterized as memes. These situations show how a memetic account of culture can explain why some cultural traits are more likely than others to mutate; how their mutation rate may be influenced by external events; why some directions of mutation are more likely than others; and how the processes by which cultural information produces effects on the world (or in the language of memes "is translated into the phenotype") will also limit the mutations that are able to occur in practice.

In the light of the genetic considerations I have outlined, we should expect memetic mutation to display no intrinsic bias towards increased fitness – and indeed it is the case that the mistakes made in learning or carrying out cultural activities display no intrinsic bias towards improvement. Imagine, for instance, that you are learning how to make chocolate chip cookies. If you are tired and not paying much attention, then you might mix in an ounce of salt instead of sugar. If you do not check your stocks of ingredients before you start, then you might find that you haven't enough dark chocolate and end up using a mixture of dark and white chocolate chips in the cookies. An ounce of salt would render the cookies inedible, and this mutation would not last long; certainly it would not be replicated (this is not to say that it will never happen again, but of course repetition is not the same as replication – and the ghastly results of its first occurrence will make it far less likely to happen a second time). If, however, you use a mixture of chocolates, then you may even prefer the resultant cookies, and in this case the mutation may be replicated and even established as your normal practice (as well as that of your children, or of anyone else whom you teach to bake). Neither mutation appears to be *intrinsically* more likely: in practice, its likelihood will depend on the environment – on the similarity between the ingredients' containers, for instance, or the level of your baking experience.

Like genes', then, we should expect that memes' mutation rates may be influenced by events external to themselves. Moreover, since culture is so vast and diverse, it is reasonable to expect that the intrinsic tendency towards mutation will also vary from meme to meme. A brief consideration of what actually goes on in culture reveals that this is indeed the case. The stability of well-established, successful cake recipes, for instance, can be contrasted with the frequent changes in fashionable clothing. Similarly, some directions of mutation are more likely than others: you are more likely to replace sugar with salt than with flour; next season's fashions are less likely to resemble last season's than to differ from them.

Finally, in the genetic case we saw that mutations were limited by the process of translation from genes to phenotypic effects. Is this the case for culture: can there, in other words, be copying errors (memetic mutations) that are unable to produce variation at the practical (phenotypic) level? This is in fact the most significant way in which variation might be limited, and is clearly to be observed in culture. An architect can draw gravity-defying buildings until he is blue in the face, but he will not be able to have them built; a composer can write music for the violin involving notes lower than its bottom G (198 Hz), but no properly tuned violin will be able to produce that music. There are limits to which of memes' mutations will be available to selection – limits which depend as much on the previously evolved memetic environment as on the laws of nature – and this can be seen in the gulf that sometimes divides brainwaves from their implementation.

Memetic Innovation as a Mental Process

All of these examples appear at first sight to be uncontroversial, but there is an obvious objection to the claim that cultural variation displays no intrinsic bias towards fitness. Unlike genetic mutation, which is an essentially mindless process, cultural changes may be directed by intentional human decisions – and this must surely make them more biased towards fitness than their genetic counterparts. Whereas genetic variation is so broad that "it relies primarily on massive parallelism rather than strategy," in culture it seems that "variations are generated strategically,"[3] rather than randomly, and will build on what has gone before.

One response to this criticism is to claim that any apparent dichotomy between "parallelism" for genes and "strategy" for memes is illusory. On this view, memes depend on human minds for reproduction, innovation and selection, and any "bias" towards fitness is simply the consequence of human minds' ability to think swiftly through the alternatives before coming up with a particular variation. Thus memes do generate similar proportions of useless and useful variations as genes, and memetic variation often *is* a process of parallelism rather than strategy – it's just a parallelism generated by an astonishingly swift, internal and frequently unconscious mental process.

I think that there is a lot of truth in this response, but nonetheless it does not quite succeed in turning aside the original objection, which touches on an issue of great controversy within memetics: the relationship between memes and the human mind. In particular, is our intelligent consciousness a problem for the claim that culture develops via an *unconscious* evolutionary algorithm? Or indeed – as some would claim – is the problem actually the other way around, with meme theory undermining our old assumptions about our identity as intelligent and autonomous persons? This debate would involve rather a detour from our present concern with memetic variation, but I return to it in more detail in Chapter 12. For now, suffice it to say my contention remains that a memetic account of culture is at the very least consistent with observations of how mutations occur when cultural information is copied.

Innovation and Memes: Recombination

If we anticipate that memetic innovation may also be due to recombination, then what does that lead us to expect to observe in culture? If it is true that cultural changes can be sometimes be characterized as memetic recombination, then we should be looking for instances when cultural innovation – either at the individual level (when a person learns something new) or at a social level (when genuine novelties are discovered or created) – results from the collision between existing knowledge and skills.

Certainly, teachers make use of this technique in order to convey information to their pupils. A good teacher will start from the point that her pupils have already reached: she will try to bring their existing knowledge into a novel environment, which may bring out previously unsuspected implications of holding that knowledge, or even facilitate the acquisition of new facts and skills. The result of bringing old ways of thought to a fresh situation is to force their recombination, and this in turn produces new ways of thought and new knowledge. In History, for example, a teacher might start by exploring students' existing knowledge about human relationships and how people tend to react in certain situations; if she next draws their attention to the historical events they have studied so far, then, by using this as a novel context for their understanding of human interactions, she will be able to help the pupils to explain and even predict the events that followed. Just as meme theory would forecast, a truly useful way of leading someone to new information is to provide the opportunity for the recombination of aspects of his existing knowledge.

The same technique can enable pupils to learn *from*, as well as *about* the facts that they are studying. Teachers of Religious Education know that this method is far more effective than simply delivering the bald facts: you can inform pupils that Christians believe that Jesus taught important lessons in parables, have them investigate the content and discover the possible meaning of some particular parables; or you can do all that and then encourage them to reconvey the same meaning in a more modern story – and they cannot achieve this without truly understanding that meaning, from which they can take away valuable lessons for themselves. This need not be proselytizing by the back door. Students who do not wish to take on board the Christian message can, nonetheless, learn other lessons from studying the parables, none of which need be explicitly stated by the teacher: the aim of good RE is to give the pupils opportunities to take away from the subject what is appropriate to their individual situations, rather than to impart the teacher's own moral or social views. Thus each pupil might well learn something different from the same lesson: all have brought different experiences and knowledge to the lesson, and that (as predicted by meme theory) will shape their responses to what they hear.

Nor should it make much difference whether a person is led to such recombinations by the guiding hand of a teacher, or by mulling things over in his private thoughts: perhaps a new external environment provokes a novel way of looking at existing knowledge or of tackling a familiar task; or perhaps meandering trains of thought bring about the collision.

Either way, the results could be the same as if the pupil had been guided to the new environment by his teacher. Existing elements of his knowledge, which had previously been inactive, will be stimulated. Putting it in terms of memetics, the forced recombination of his existing memes will reveal which of them is relevant to the novel situation, and thus new memes may be acquired.

Philosophers' "thought experiments" (in which the imagination is exercised in a controlled fashion in order to examine theoretical implications or to explore conceptual boundaries) seem to be especially pertinent to this picture,[4] since their very purpose is to enable us to access or re-represent familiar information. In other words, reconceptualization or re-representation may helpfully be seen as an example of recombinative innovation in memes.

Memes and Their Alleles

What of the recombinative *restrictions* that were noted in the genetic picture? Not just any old bits of the parent chromosomes change places: they have to correspond exactly, in that a gene's alleles must all control the same phenotypic effect. Many memeticists assume that there will be no such corresponding restriction in the cultural world, for there is no such thing as a memetic allele. In my opinion this assumption is wrong. A replicator's alleles are at least partly *defined by* the phenotypic effect that they control: that is what makes them alleles of *that* particular replicator, providing variety amongst the effects that it controls. I see no problem with the claim that there may be a variety of alternatives to any particular cultural trait, just as there are variety of alternatives to genetically controlled traits like eye colour and height.

Such memetic alleles will have effects that correspond to the meme in question: you could replace the meme for a fence with one for a hedge but not with one for a bike; and you could replace the meme for a bike with one for a tricycle but not with one for a hedge. As Liane Gabora puts it, "When considering the problem of having to get out of your car every day to open to garage door, you would not think about doilies or existentialism, but concepts related to the problem"[5] – and this is explained by the fact that neither a doily nor existentialism is an allele for a method of opening garage doors.

Nonetheless, there is a theoretical – rather than merely observational – objection to the concept of memetic alleles, which has been put forward by Dan Sperber. He argues that it is a mistake to try to abstract common properties from a group of cultural tokens, calling that abstract description a "meme" and the alternative concrete versions its "alleles". When studying cultural phenomena like myths, for example, he says that it is more useful to concentrate on "the many public and mental versions and their causal chains",[6] than to seek to abstract *one* canonical version from the similar myths that occur across different cultures. He points out that the myth's canonical version would not actually exist anywhere, any more than the abstract version of a cross-cultural concept like "marriage" can actually be found in any one of the cultures where similar practices are observed. So how can the abstract representation be of use in explaining the cultural facts that we are trying to study? In his opinion, the temptation to put groups of cultural phenomena under one label (a meme and its alleles) stems from a tendency to "exaggerate the similarity of cultural tokens".[7]

My own view is that the stark choice presented by Sperber, between an abstract representation and concrete cultural tokens, is a false dichotomy. What is it that makes m_1 a token of the same meme as m_2? They are both m-ish memes, on my account, because both are alleles of the same meme – and surely there is no more harm in attempting to give an abstract definition of the "typical" content of m-ish memes than in trying to identify the sorts of genes that appear at a particular locus on a chromosome. Although no one has the abstract property "eye colour", still that is a good approximation of the sorts of alleles that appear at a certain genetic locus, and the abstract definition will help you to identify the relevant alleles. It helps you to link a group of genes with the correct range of phenotypic effects. Similarly, although neither the canonical version of a myth, nor marriage as characterized in an abstract definition will be instantiated in any culture, still those abstractions can serve a useful purpose: the canonical version of a myth does help you to find the alleles of that sort of myth, and the abstract definition of marriage will help you to identify which practices will fit into that category.

I suspect that the problem arises because of a temptation to label the abstract representation "the meme", and its concrete versions "the alleles", mistakenly seeking a default or master copy of a meme, in a way that we don't for genes. Rather, there may be many alternative versions of a portion of cultural information, producing a variety of versions of the effects that it controls – just as there may be many alternative versions of a portion of chromosomal material, producing a variety of versions of the effects that it controls. The link between all of these versions (in either case, cultural or biological) is that all control the same phenotypic effect, which can usefully be identified with the help of what Sperber would call an "abstract synthetic version"[8] of the relevant information.

One significant difference from the genetic situation should, however, be noted. In language, as Steven Pinker[9] has pointed out, the possibilities of innovation due to recombination are much greater than they could ever be in DNA, since there is no determinate message length – no fixed number of chromosomes. The question whether memes might be realized in language has not yet been addressed, but memes *are* discrete representations, so whatever their basis, it may well be described as one of Pinker's "discrete combinatorial systems" (i.e., a system that is rule governed, containing a finite number of recombinable discrete elements). Since this is so, recombination may be a more significant method of memetic innovation than it is for genes.

Too Much Variation?

One final issue regarding the subject of memetic variation concerns the problems that this area of cultural evolution might create for meme theory in general. The philosopher Daniel Dennett[10] has asked whether "one of the hallmarks of cultural evolution and transmission [is] the extraordinarily high rate of mutation and recombination", pointing out that "evolution goes haywire"[11] when mutation rates are too high. Fortunately for meme theory, he does not comment on whether he regards the rate of memetic variation as high *in relation to the rate of memetic replication*, which is the relevant question – and it has already been noted (by Dennett as well as myself and others) that memetic evolution is in general a much swifter process all round than its biological counterpart.

Conclusions

Processes that might usefully be characterized as memetic mutation and the recombination of memetic alleles can, it has been demonstrated, be observed in the growth and change of human culture. Indeed, such a characterization provides a useful perspective on these processes, explaining to some extent what is going on behind the apparent unpredictability of mistakes, alterations and novelties that pepper our everyday lives.

6

Selection

Evolution depends on selection as much as on replication and innovation: if all novelties had an equal chance of success then there would be no gradual development, in culture any more than in nature. The information that is preserved and copied in cultural traits must, in order for evolution to occur, be subject to some sort of struggle for survival. Putting this another way, meme theory needs some sort of criterion of memetic "fitness" – something that ideas and skills have in common, in virtue of which their relative success rates can be subject to systematic study – for without this it breaks down into the trivial statement that out of many new ideas and skills some survive whilst others do not.

Glancing back to the natural world, we know that the general fitness criterion for genes is the influence they have on an organism's longevity and fertility – its ability to find food and sex – and that this will be affected not only by the genes that it possesses but also by the rest of the gene pool and the external environment.

So what is the general fitness criterion for the cultural population of traditions, ideas, tunes and designs? It makes sense to say that, as for genes, memetic success will depend on three separate factors: the content of the meme itself; the way in which it fits with other memes; and the external environment – the minds and surroundings of the people whose attention it is trying to attain. There is a struggle for existence because a vast array of memes is competing for the limited resource of human attention, and therefore the fitness of any given meme will be influenced chiefly by its ability to gain and retain attention. Gaining someone's attention is a means of replicating itself (cf. sex), and retaining that attention is a means of prolonging its survival (cf. food).

Notice that "attention" in this context need not involve a possessor's constant active awareness of the meme. We are obviously not aware in this way of "hard copies" of information, such as that printed in books, but nonetheless their preservation ensures their ongoing potential to grab someone's attention, and thereby provides an efficient method of survival.

Having identified this very general fitness criterion for culture, we can now move on to more focused questions about particular memes. How can an individual meme, given the specific demands of its environment, ensure greater access to its means of reproduction and survival (i.e., gaining and retaining attention)? Obviously the answer to this question, given in terms of the interaction between the meme's environment and its phenotypic effects, will vary widely between cultural contexts. If the meme is a scientific theory then it must enjoy some degree of explanatory success, must not contradict existing theories but must also accord with the available perceptual evidence; if it is a melody or a picture then it must be aesthetically pleasing, and the conditions that this entails may be determined by time or place; if it is a recipe then it must result in good tasting, nontoxic things to eat, and again decisions about the former may differ between cultures.

Between the two extremes, of the general truth that the successful memes will be the ones that gain and retain lots of people's attention, and the particular facts about how individual memes can do that in widely differing cultural areas, there must also be general factors that affect memes' fitness across particular cultural areas: memes may be "locally relevant, and hence culturally successful, in part for universal reasons".[1] What are the most significant factors at work in cultural selection?

Factors in Memetic Selection

The previous chapter's discussion of "recessive" memes forms an important backdrop to the discussion of cultural fitness. There the difference was noted between acquiring information and taking it on board: we all possess a fair amount of knowledge about beliefs and theories to which we ourselves do not subscribe. Such information can be characterized as recessive, in that we are capable of passing it on to others, but it has little or no effect on our thoughts or behaviour. It is therefore doubly hampered in the struggle for survival, being less likely either to be remembered or to be replicated, even if we do remember it, than information that we accept as valid. In the discussion that follows, therefore, the assumption is largely that the success of a meme is bound up with its acceptance by those who acquire it – but it should be noted throughout that this is not always the case, and that some memes may grab a great deal of attention purely as a result of their novel or shocking content, even though their possessors do not for one minute believe them. Examples might include outrageous urban myths, or bizarre-sounding scientific theories to which previous generations subscribed, both of which can sometimes survive on the back of their novelty value.

The Memetic Environment

Bearing this in mind, we return now to the general factors affecting memetic fitness. One of the most significant aspects of any meme's environment will be the other memes that are present in that culture, and this is the factor that will often dominate the fate of novel memes in particular. In order to be accepted, an idea has (usually) to be compatible with those already in existence – which means that selection will favour memes that are capable of exploiting the current cultural environment. The result will be coadapted meme complexes which bestow further benefits on their members in addition to the initial privilege of admission: as the complexes grow in size and strength, they will become more difficult to penetrate, providing protection against invading, contradictory ideas. This is analogous to the complexes of coadapted genes to be found within particular species, and typically we should expect to find protective meme complexes within specific cultures. It also reflects a fact to which the previous chapter pointed: that the direction of evolution will be dependent upon what already happens to exist. For specific novel replicators – both genes and memes – this will mean that their success or failure will be partly determined by the prior existence of other replicators in their area. "Much as the evolution of rabbits created ecological niches for species that eat them and parasitize them, the invention of cars created cultural niches for gas stations, seat belts, and garage door openers."[2]

Novel genes that are incompatible with existing genes will be destroyed because they cause the destruction of their mutual "survival machine". For memes, however, the effect on the human organism is not so drastic: if someone favours a novel meme over those of her existing memes with which it is incompatible, then the chances of this resulting in her death are low. On the other hand, there may be *social* advantages in her sticking with her existing memes, since people who bend to the prevailing *Zeitgeist*, and are reluctant to resist popular opinion, often do best socially. There are other advantages, too: the tendency to favour what exists over novelties creates a positive feedback loop with the tendency to build stable complexes.

One reason for this may be that the amount of effort already invested in acquiring a meme will have been entirely wasted if, whenever an alternative is encountered, the original stands as great a chance of being rejected as the novel competitor. Rather, as soon as someone has decided that one meme is worthy of his prolonged attention, a tendency to favour it would be advantageous: instead of assuming (I don't mean consciously) that a new meme is as likely to be the correct choice as the old one, it is much more efficient for him to work on the unconscious assumption that his existing memes would not have been acquired were they not worthy of his attention. He should only acquire a novel meme if it either is compatible with the old ones or has obvious enough advantages over them to compensate for his previous investment. Such a tendency to build on what already exists would lead to stable meme assemblies, and at that point any incoming meme which contradicts one of the assembly's elements faces even greater opposition. Rejecting the existing meme now entails rejecting the whole assembly; conversely, the incoming meme now needs to have obvious advantages over a whole complex of existing memes. Thus the very existence of the assembly increases the advantage of sticking with the existing memes, and that process in turn builds up the assembly.

Another reason for conservatism may simply be that any meme which is invasively strong enough to secure attention in the first place already enjoys a certain amount of success. Presumably if it can gain attention over existing memes (which are also relatively successful), then it will also be able to retain attention over most newcomers.

On the other hand, this will depend on the *amount* of attention that it has gained, as compared with the amount that the newcomer is potentially able to gain. In other words, the fitness of a novel meme for an existing assembled complex will depend not only on the memes that are already within the complex but also on the commitment with which those memes are held. A new meme that contradicts an idea which is not so fundamental to the assembly – one that is not too deeply entrenched – has a better chance of success than a novelty that is in conflict with a "keystone" meme. Memetic fitness depends on the ability to gain and retain attention, and it will be much harder for a meme to do this if it contradicts existing, deeply entrenched alternatives. A meme that demands I turn my back on a belief or skill that I hold dear or have learned to trust will not easily persuade me that it is worthy of my prolonged attention. There is more on this later in the chapter.

The Physical Environment

So in the struggle for existence memes are selected via their phenotypic effects, which must be compatible with (i.e., able to penetrate) the existing meme assembly in its cultural "area". Yet this is not the whole story. Memes' success is not simply a matter of their effects' compatibility with the existing *cultural* environment: they must also be anchored to reality by according with the perceptual evidence. The reason why the "flat earth" meme would not succeed today is not just that it is incompatible with existing *theory* but also that it is contradicted by the best available *evidence.*

No matter how much potential a meme has for longevity and fecundity, this will never be realized in some physical environments. The meme for riding a bicycle to work will be seen as much more attention worthy in the calm flatlands of Cambridge than amongst the weather-beaten gradients of the Yorkshire Dales; the fashion meme for miniskirts in Siberia, or for Aran sweaters in Florida, is not likely to retain much attention. The importance of the physical environment for memetic fitness is apparent in almost every cultural area. Like genes, memes do not succeed or fail per se. As a gene is dependent on a coadapted gene complex for protection, and on surroundings that are kind to the phenotypic effects it produces, so a meme needs a receptive cultural environment *and* an external world that accords with its effects. Without these conditions, it will never have the chance to be fecund.

The Genetic Environment

If memes' success can be determined by their memetic and physical environments, then what of their genetic surroundings: is memetic fitness ever determined by human biology? A previous section on sociobiology (Chapter 2) rejected the suggestion that Darwinism could account for the diverse details of culture, chiefly on the grounds that the rate of cultural change just could not be picked up at the genetic level. On these grounds, we should expect that the broad facts about social structure will be such as to raise our biological fitness, but that the majority of uniquely human traits will be determined by the constraints of *cultural* fitness. Nonetheless, it would be odd if our biological ancestry did not exert some influence on our culture – if only to the basic extent that, for instance, the meme for binding the whole of the Oxford English Dictionary into one huge, unmanageable volume would be doomed to failure because we are not physically large enough to handle it. More than this, we should expect our biology to have some influence on what we deem worthy of prolonged attention. A theory that entirely removed behaviour from genetic control would not be very plausible.

It may well be, for instance, that the human genotype will provide innate biases or parameters for change which sometimes constrain the particular sequences of memetic acquisition and evolution. Thus, although the primary criterion for a cultural replicator must be its compatibility with the existing culture, still there will be occasions when its success or failure will be affected by the biological nature of its possessors. So, for example, whilst it is true to say that the popularity of films will reflect the culture in which they are shown (big hits in the United States may flop in the United Kingdom), it will also be the case that their reception is affected by cross-cultural, biological factors (e.g., the appeal of love stories, action adventures, or films about large-eyed, deep-foreheaded creatures such as ET). Memetic fitness criteria are, then, sometimes determined by our genotype. This is not to say that, even in these instances, memes' effects on our *biological* survival will determine their fitness (note that what we were selected to prefer in the distant past may not now even affect, never mind raise our biological fitness). It is merely to point out that memes' own success rates will sometimes be influenced by our biological nature.

Human Psychology

An important factor in our biology is, of course, our psychology. Indeed, some would go so far as to say that the nature of the mind (as opposed to the nature of a meme's own content) is the *prime* factor in determining a meme's success or failure,[3] and others that there is an even closer connection between memes and psychology than this: that at least some emotions *are* memes, and that in some forms of social interaction those memes can be copied between different people.[4] Empathy, for instance, might be a form of memetic transmission of emotions: you feel sad when you read in the newspaper about strangers' tragedies, because you can empathize with their situation; you feel, for a brief time, the *same* emotions as they do. Another case of emotions as memes might be the transference of feelings between toddlers or teenagers and their parents, when the child's strong emotions are picked up and carried by their parents (or indeed vice versa). So, for example, a parent feels confused and at the end of her tether when dealing with a wilful two-year-old, because she is unconsciously picking up the confusion and panic of which the child's tantrum is a display.

According to my own characterization of cultural evolution, however, emotions do not fit into the category of memes, for the key reason that they are non-representational. If memes are representations of cultural information then emotions are not memes, for our feelings do not carry information *about* anything, unlike our thoughts. Our feelings are of course provoked by situations, and indeed a standard therapeutic question regarding our emotions is, "What's that about?" – but the meaning behind that question is not, "What information is represented by your anger (or tears, or whatever)?" but rather, "What lies behind the strength of your feelings about this?" or, "What is causing you to feel like this?"

The situation is somewhat confused by the loose way in which we often talk. It is not uncommon for people to say something like "I feel that our relationship is over", where despite beginning with "I feel", such sentences express a thought rather than a feeling – as demonstrated by the fact that the underlying emotions are actually left unsaid by such statements; does the speaker feel sad about the end of the relationship, or angry, relieved or what? Such imprecise language is not, however, an accurate guide to the true state of affairs: there is a real distinction between thoughts and feelings, and therapists are trained to raise their clients' awareness of this difference, so that they can gain clarity about the emotions that underlie their statements.[5]

Emotions, then, do not carry information: they are not memes but potential reactions *to* memes – as well, of course, as more direct reactions to situations, people, memories, thoughts and so on. The emotions that a meme tends to provoke might help in its selection, as when we are stirred or touched by a piece of literature, art or music, which we consequently retain in our memories and/or recommend to others. The feelings provoked may even be almost invariant between the people who encounter it, but even so what is being copied in such cases is not the emotion, but the content of the piece (to which the emotion is a reaction). Returning to the example of empathy, it is on this account the result of our imagining ourselves in the relevant situation, and responding to it in the same way as the person who is actually undergoing that experience: we are not so much copying their feelings as reinventing them for ourselves. Similarly, parents whose emotions mirror their offspring's anger or confusion are not imitating those feelings, but rather misattributing their source.

Yet human psychology obviously *is* a crucial factor in directing meme selection. Human minds process and filter certain types of knowledge differently from others, and memetic success or failure will be affected by this as well as by other environmental factors. An individual may be more likely to accept a particular meme, for instance, if it is shared by someone whom he admires, or if it has been adopted by the majority of his social group. We *trust* certain sources of information – our parents when we are small, or people whom we accept as experts in a field of which we ourselves are rather ignorant – and are more likely to accept information when it comes from these sources, even if our own understanding of it remains rather hazy. Unlike facts about the natural world, which we automatically and stringently test for consistency with each other, we may subject facts about more complex things like science or religion to less stringent testing. If some seem inconsistent with others then we are open to the possibility that it is our understanding, rather than the concepts involved, which are at fault. (The lay person's knowledge of physics, for example, may include the fact that tables are made up of many whizzing atoms – with a tacit mental footnote "whatever atoms are".) It is this which enables us to accept unclear or mysterious claims on the basis of trusted authority: we may check out their consistency with our assumption about the trustworthiness of that authority rather than with other facts or external evidence.[6]

It is therefore clear that human psychology is of some relevance to memes' relative success rates: "every individual differs in his or her susceptibility to adopting particular memes depending on genotype, development, individual experience and social environment, and this susceptibility is not itself exclusively the product of past meme adoption."[7] The direction of cultural evolution will be influenced by the characteristics of both memes themselves *and* the human mind – and Chapter 12 explores in more detail the nature of the interaction between the two.

Memetic Content

It is interesting to make the distinction between the environmental influences (whether cultural, physical or biological) on memetic success, and the ways in which the meme's own content might influence its survival and replication rate. The content of genes, for example, carries an inbuilt means of ensuring success: you cannot acquire a gene without acquiring also the instructions to replicate it, for a vital part of DNA's function is to make copies of itself. In what ways might memes' content affect their fitness?

On the whole, memetic replication techniques seem less aggressive than DNA's: few memes carry instructions for their own replication. Yet what if a meme did contain, as part of its content, the instruction to make copies of it, so that part of its executive effect were its own replication? It would have a built-in advantage over memes that have to rely on the usual processes of imitation and communication to make copies of themselves – not because it was automatically more able to gain (or even retain) our attention, but because it would have *guaranteed* its replication whenever it *was* acquired.

In some cases memes do seem to have achieved this: Dawkins has referred to religions as "duplicate me" programmes (see Chapter 8 for discussion of his suggestion), and it is certainly true of many religions that they have an evangelical element. Of course it begs the question of their truth to say that they are acquired (i.e., accepted as true) by so many people simply because "replicate me" is part of their content: it may be that "replicate me" is part of their content because they make true claims about matters that carry such weight with us. There are, though, less controversial examples to be found. Consider for instance the folk-song tradition, which contains as a vital element of its content the claim that the songs are important parts of our heritage, and should therefore be preserved by one generation for the next. The possession of some political views (usually the more extreme versions) also entails the demand that they should be held by all.

So, although memes are usually less aggressive than genes, it appears that some do include the instruction to make copies of themselves. Why, given that they have this innate advantage, have such memes not swamped the meme pool? The answer is twofold. First, even if you acquire a meme that inclines you to replicate it, your success in so doing will depend on the rest of the meme pool – and if that is unreceptive to your meme then your attempts to make copies of it will fail. Secondly, and as a corollary to this, a meme's possession of the "duplicate me" factor will not automatically imply that it can thereby bypass normal rationality. Flat-earthers could be as evangelical as they liked; in the twenty-first century the "flat-earth" meme would have little success.* The mere fact that a meme has "replicate me" as part of its content does not mean that it is *not* worth replicating – but nor will it automatically increase a meme's ability to grab and keep our attention. Thus the external and intermemetic context of any meme will still be the most significant factor in its selection, just as it is for genes.

Relative Fitness

It has been suggested, however, that there is a significant difference between genes and memes when it comes to the *relationship* between their

* Although see Chapter 5's discussion of recessive memes for a corollary to this statement.

content and environment. Genetic success is intimately related to the actual fit of genes' phenotypic effects to their physical environment, whereas in some cases a meme's success may be guaranteed by the *perceived* fit of its effects to the environment.[8] What this means is that, whereas a gene must succeed or fail in virtue of the effects that it actually exerts, a meme may succeed even if it does not have the effect popularly attributed to it, or even if the idea that it represents is not true.

Thus, for example, someone might build a car in a particular shape because he *thinks* that it will increase the car's speed, even though in fact it will not – and that car may sell well because the public shares his misconception, thus perpetuating the meme. Or a political party may be voted back into power on the strength of the claims that it makes about its achievements in reducing taxation – and supported even by those people who are actually paying more taxes overall, without realising it. Memetic success, unlike genetic success, is not necessarily linked to reality.

In fact this apparent distinction between the two realms of selection may more helpfully be viewed as an interesting parallel. Again it will prove fruitful to explore the biological case in a little more detail first. In nature, a feature will be selected because of its fit to the *actual* world, so of course the fitness itself must also be actual. Since forward planning is not possible in nature, biological fitness must always be determined by the organism's immediate needs. Even if a characteristic might be helpful to a species in the long run, the only factor that will determine its success will be its immediate effect on the organism's ability to gain food or sex – and this will depend not only on the current environment (what does the organism need, and what resources are available?), but also on the existing gene pool (with whose members any novel genes will have to compete). Genetic fitness is, in other words, a relative concept: what gives a selectional advantage in one particular time and setting would not necessarily have done so in different circumstances.

In culture, a parallel story can be told. Features will often be selected because of their fit to the *mental* world, which may itself be hypothetical (i.e., not linked to reality). Since absolute knowledge is not possible for humans, cultural fitness must always be determined by our current *perception* of appropriateness. Even if a new idea or design might prove truer or more useful in the long run, the only factor that will determine its success will be its immediate ability to gain and retain attention – and this will depend on both the current environment (what do people currently perceive as appropriate and desirable?) and the existing meme pool (with whose members any novel memes will have to compete). Like genetic fitness, then, memetic fitness is a relative concept: what gives a selectional advantage in one particular time and setting would not necessarily have done so in different circumstances. In both cases the fit of a replicator's content to its environment is relative to what already exists – either in the replicator pool or in the environment – and is not absolute.

Conclusions

A meme's own content may, then, be a fairly arbitrary factor in determining its success: its fortune in the struggle for survival will always be relative to context. As memes struggle to gain and retain the attention of human minds, their success or failure is in this sense influenced more by the environment than by their own content. Novel memes must be fit for the existing body of culture, for the physical environment and for the dictates of human biology and psychology, in order to stand a chance of being copied accurately or enduringly.

7

The Story So Far

Memetics must be able to provide a convincing account of how the three essential elements of evolutionary theory – selection, variation and replication – work in culture. The preceding four chapters have taken on this challenge but have in the process raised a variety of questions on which the credibility of meme theory is equally dependent.

Selection

"Selection" means that some replicators are favoured, survive and propagate, while others fail and become extinct. Genes are selected via their phenotypic effects, and the evidence for such selection is therefore to be sought at the level of the phenotype. Nor is it hard to find. An abundance of extant and extinct species – living organisms, creatures that have been wiped out within living memory, and fossil records – all contribute towards the plausibility of natural selection in biology.

If memes, like genes, are selected via their phenotypic effects, then it is at the phenotypic level that we must search for the evidence for their selection, too. Again, there is plenty of evidence for selection in culture: theories, tunes and methods that are popular at present; ideas that have been rejected within living memory; written records of the theories, fashions, skills and music of past generations, all demonstrate the differential survival of certain areas of culture.

The previous chapter's discussion provides, in addition, some theoretical insight into the selectional pressures on memes: the limited capacity and attention span of human brains; assembling compatibility pressures; a variety of constraints specific to the different cultural areas; the physical, genetic, memetic and psychological environment; to a certain extent the content of the memes themselves. Most significantly, it emerged that memetic selection will depend on memes' respective abilities to gain and retain our attention *in the current context*: fitness is always a relative concept.

In the struggle to be selected most memes do not, unlike genes, automatically come with instructions for their own replication, but Richard Dawkins has raised the possibility that at least some cultural elements do arrive so equipped. These he has labelled "viruses of the mind", and the next chapter discusses this claim – a discussion whose implications will reveal much about the true nature of memes.

Replication

A second evolutionary process is replication. Genes replicate via meiosis or mitosis, which preserves the information of their constituent DNA. They have very different properties from those of their "survival machines": genes exist and function in their own right, distinct from the phenotypic results via which they are selected, which exist and function at a more composite level of the evolutionary assembly.

Memes, too, have been characterized as existing and functioning autonomously, and I have claimed that they owe their distinct properties to the representational content in which they consist. That content must be preserved in such a form as to be available for activation – it must constantly have the potential to give phenotypic results – but, like a gene, it may be recessive and sometimes produce no such results. The results that it does produce must be generally applicable in a variety of contexts. An adequate theory of representational content can explain how memes are able to fulfil these roles, in the same way that a proper understanding of DNA revealed the mechanisms of genetic heredity. Memes have their basis in representational content, just as genes have theirs in DNA. This addresses the question of *what* memes are, but a theme that recurs throughout the remainder of this book is the question of *where* memes are to be found. In other words, how is their representational content physically realized?

It is obvious that a key feature of memetic content must be its replicability; without this property no representation could be a meme. In practice, the transmission of memetic content will be facilitated by such standard cultural methods as imitation, teaching and everyday communication. There will be constraints, however, upon which sorts of transmission methods will be able to support memetic replication, since none will suffice which cannot account for the exponential growth of cultural complexity – and any replication of complex information depends on a process of hierarchical assembly.

Memetic replication must, in addition, be dependent on the human ability for social learning, and Chapter 4 raised questions about the types of social learning that can support memetic replication. In particular, there is a dispute within memetics about the significance of imitation for cultural evolution. More fundamentally than this, there is a debate to be had about whether the transmission of cultural information involves *replication*, as such, at all. Again these crucial questions are tackled in the second half of this book.

Variation

The third aspect of evolution is variation, and in Chapter 5 I argued that mutation and recombination – the methods of genetic variation – also provide a good account of the ways in which cultural variations might arise. Memetic mutations will be subject to certain biases and limitations, determined by what already exists and its assembled organization. The recombination of memetic alleles may well be the more usual method by which cultural variations arise, and I defended the view that memes do, indeed, have alleles.

An outstanding issue, however, is the fundamental question whether cultural change is really facilitated by *particulate* units of selection, in the way that biological developments are. I shall argue that it is.

Memes and the Mind

A recurring theme throughout almost all of these areas of discussion has been the question of the relationship between memes and the mind. Memes are selected by virtue of their fit to the cultural environment: what part do humans play in directing their selection? If memes are preserved as representational content, then where does that content stem from, and where is it stored? If memes are transmitted via a variety of means of communication, then to what extent are human agents necessary and/or sufficient elements of these copying processes? Memes must vary if cultural evolution is to happen: is this a passive event, or are humans actively involved in it?

According to some of the most respected and vocal memeticists, such questions are not worth answering; the reality is that there is no significant distinction between memes and the human mind. Others would rather reject the meme hypothesis altogether than allow it such free reign over our theories of mind. It is time to set this book in its own cultural context, by investigating some other writers' views of memes.

8

The Human Mind: Meme Complex with a Virus?

Since Richard Dawkins first proposed his meme theory in 1976 there have been a number of attempts to develop and defend it, as well as some rather misplaced criticisms.[1] In this chapter and the next, I explore several such commentaries. Chapter 9 focuses largely on the issue of imitation, as discussed by Susan Blackmore, Dan Sperber, Robert Boyd and Peter J. Richerson. Here I examine two of the best-known applications of memetics: Dawkins's own attempt to embrace viruses within the cultural side of the analogy, and Daniel Dennett's claim – one of the most significant that has been made for the potency of meme theory – that memetics can explain the emergence of human consciousness.

Richard Dawkins

Richard Dawkins has speculated about the extent to which a certain type of cultural replicator might be seen more as the analogue of a virus than of a gene.[2] Famously, as an example of this sort of replicator he uses religion, and concludes that it "is best understood as an infectious disease of the mind".[3] This analysis he uses to add weight to his already well-publicized conviction that truth is incompatible with religion, for it implies that large sections of the human race are even now devoting themselves, not to the service of God, but to the propagation of a virus. Since Dawkins regards biological evolution as an alternative to God, it is perhaps not surprising that he should also use his theory of cultural evolution to explain away religious belief. This section investigates whether his arguments constitute a valid application of meme theory.

Good Memes Versus Mental Viruses

Dawkins starts with a discussion of the nature of physical viruses, and describes them simply as "Duplicate Me" programs written in the language of the DNA code. Their advent was "inevitable" once the cellular machinery for DNA replication had developed, for such an apparatus provides the perfect niche for subversive parasites which hijack genes' replicative machinery in order to make copies of themselves. Thus Dawkins contrasts viral with genetic methods of survival. Both use the same means of replication, but genes produce it and viruses hijack it: genes' replicative success depends on their producing beneficial effects on their possessors' chances of survival and reproduction, whereas viruses' replicative success depends merely on their ensuring that they are replicated. (Compare the fact that a robin's successful procreation depends on its *providing* a safe environment for its chicks, in which they are continually fed; a cuckoo's depends merely on ensuring that its chicks *enter* such an environment.) Raising the possibility of cultural viruses, Dawkins says that our brains, with their naturally selected openness to memes, also provide an environment that is ripe for parasitic exploitation.

If we accept Dawkins's hypothesis, then in what way should we expect "mental viruses" to differ from memes? To uphold the biological analogy, we should expect mental viruses to succeed by parasitizing the normal process of cultural replication, in order to make copies of themselves. We should expect the sorts of ideas that are viral rather than memetic to be successful not because of any replicative advantage they hold over other ideas (are better predictors or more aesthetically pleasing, etc.), but merely because they have found some way of ensuring that they are transmitted. Putting this another way, whether a given replicator is a virus or a meme will be determined by its method of replication ("normal" or parasitic), rather than by its content.

Dawkins begins by explaining the properties that make a medium vulnerable to parasitic exploitation: its (almost) accurate replicative powers, and a willingness to obey the instructions that it is replicating. The human brain, he argues, has just these properties. He points out that, in addition, the mind of a *child* is especially susceptible to parasitic exploitation: it needs to be receptive to new ideas, in order to soak up a whole culture and language.

In this context, Dawkins first introduces the idea that religion is just such a virus of the mind, able to manipulate the thought processes of its victim, although that victim will be unaware of being so manipulated. Later he adds that the religion virus incorporates certain features that make sure that it sticks and spreads. It includes, for example, the ideas that faith (which Dawkins characterizes as belief without evidence) is a virtue, and that even if you lose your faith, you should teach it to your children in order to give them the choice of believing or not. "Religious doctrines survive because they are told to children at a susceptible age and the children therefore see to it, when they grow up, that their own children are told the same thing."[4]

Dawkins contrasts such "viruses" with "good" memes. He says that we should be careful not to apply the viral analogy to *all* ideas and *all* aspects of culture: some are more like "good genes" than self-serving, "Duplicate Me" viruses. Indeed, "Great ideas and great music spread, not because they embody instructions, slavishly carried out, but because they are great. The works of Darwin and Bach are not viruses."[5] Equally, although scientific ideas might seem to spread epidemiologically, in fact they spread "because people evaluate them, recommend them and pass them on".[6]

Science: Meme or Virus?

In such statements, Dawkins comes perilously close to labelling only those things of which he approves, as "great" and nonviral. Admittedly Bach's music is enduringly popular, but at the moment there are other forms of music that are statistically *more* popular. Is this because of their viral nature, or because they too are intrinsically "great"? It may be that their fans would argue for the latter, whilst the rest of us would tend to go with the former idea – and the evidence either way would be highly speculative. Perhaps the ideas that Dawkins would – often in conjunction with most of the rest of us – wish to label "great", are actually just the ones that are most compatible with our own time, culture and available evidence. Ideas that are successful but not so apparently intrinsically valuable appear to us to be pointlessly self-replicating, but may be in the future (or have been in the past) labelled "great". More generally, it may be that the ones that are enduringly regarded as "great" are simply the ones that are compatible with the sorts of features that do not alter much between generations.

I don't want to push this thought too far, however, being reluctant to characterize as purely subjective either scientific theories or scientists' choices between them; but on the other hand Dawkins' characterization of the dissemination of scientific ideas – as spreading successfully purely as a result of accuracy and greatness – rather begs the question. It fails to account for great ideas that lie dormant for many years (e.g., Mendelian genetics), and for the spread of ideas that only seem to be great at a particular time (e.g., Lamarckian evolution). Moreover, the acceptance of scientific theses is, like religion, highly dependent on context, historical as well as geographical.

The idea of scientific theories as parasites has, in fact, been worked out in a particularly lucid account of scientific progress given by Douglas Shrader, who claims that "saying that a theory has gained general acceptance is similar to saying that a parasitic infection has reached epidemic proportions."[7] A detailed exposition of Shrader's account would not add much here, but his theory does underline the fact that the application of cultural evolutionary theories to scientific progress is, at the very least, open to interpretation. Dawkins's view of scientific theories as "good" memes, which may be contrasted with the viruses of religion, is far from being the only option.

Parasites Versus "Bad" Replicators

The crucial point, though, is that Dawkins's meme/virus distinction displays a misunderstanding of his own theory. The meme hypothesis first appeared in *The Selfish Gene*, the central theorem of which was that the gene is the unit of selection, and also "the basic unit of selfishness":[8] genes act so as to increase their own chances of survival and replication, and any selectional benefits that they confer on their survival machines (the human body, in our case), are almost incidental.

Following on from that thesis, memes were also hypothesized to be "selfish": Dawkins said that "selection favours memes that exploit their cultural environment to their own advantage."[9] Yet if this is the case, then how can the difference between memes and mental viruses rest on the question whether the replicator in question is a "good" one? In fact, neither gene nor meme theory has anything to say about the intrinsic value (i.e., "goodness") of the information that its replicators carry. As has been emphasized before, fitness is a relative concept – so we cannot just dismiss as viral those ideas whose content we see as harmful or pointless. Even for genes, it is not relevant that (usually) the virus is bad and the gene good for the individual who acquires it: both are replicators, and the distinction between them is a developmental one, a point which is emphasized by the existence of crippling and life-threatening diseases which result from the possession of a harmful gene or gene complex. The replicator-virus distinction rests on the method of replication involved, not on the content of what is replicated.

Viruses: A Biological Detail

Dawkins also seems, in his discussion of mental viruses, to have misinterpreted the very essence of the gene-meme comparison. A detail-to-detail analogy is not appropriate between genetics and memetics, and indeed the meme hypothesis loses much of its credibility as soon as one tries to claim that every feature of genetics can be carried over to culture. The point of the meme hypothesis is that if the essential features of Darwinism can be found in the cultural realm, then we should expect to see a new type of evolution taking place there. The specifics of how the new type of evolution will develop, however, need not parallel those of its predecessor, biological evolution.

It is obvious that viruses, as parasites on the biological evolutionary system, are not essential features of it. The essential features are replication, variation and selection, and genes lie at the hub of this process in the biological world. The fact that viruses also exist, taking advantage of genes' hard work, is something that has happened as a side effect rather than as a crucial element of the evolutionary system which has allowed those parasites to prosper. If there were no viruses, then evolution would still have happened (just as is the case for any other species or creature), in a way that it would not if there were no natural selection, for example. The question therefore arises whether Dawkins can salvage the virus–good replicator distinction in the cultural evolutionary system, by speculating that viruses happen to be a specific development that *has* been paralleled in it. I shall show that this is wholly implausible, as the key distinction between genes and viruses just does not arise in culture.

As emphasised, genes replicate by generating organisms, whereas viruses replicate by hijacking those gene-built organisms. The success of both sorts of replicator is affected by the effects that they produce, but genes do and viruses do not rely on *creating* the replicative mechanisms by which they produce their phenotypic effects. Genes do and viruses do not generate survival machines.

I shall argue that memes are more like viruses than genes, in that they do not generate their own survival machines. In contrast to genes – which, in conjunction with an appropriate environment, generate survival machines that may be "hijacked" by biological viruses – memes do not create the replicative mechanisms by which they produce their phenotypic effects, and thus there is nothing for a purported mental virus to hijack. If this is true, then there is no genuine analogue for viruses in culture (memes are only "more like viruses than genes" in that they do not generate survival machines – not in that they are the hijackers of some other system). Rather, Dawkins has made the mistake of overextending his meme hypothesis in an attempt to embrace what is an inessential detail of biological evolution. In the following section, I begin to make the case for this claim.

Daniel Dennett

Daniel Dennett's version of memetics is both radical in nature and crucial to his theory of consciousness. He defends the hypothesis that human "consciousness is *itself* a huge complex of memes (or more exactly, meme effects in brains)".[10] This is an ambitious claim for meme theory and one which, if true, would greatly increase its force and status. To reject Dennett's assertion that a conscious mind is the effect of memes colonising a brain is to give up a compelling pointer towards the significance and plausibility of the meme hypothesis. Apart from anything else, it is to abandon a strong and apparently attractive claim about the *location* of the units of cultural selection.

An initial reason for doubt about Dennett's thesis is that he provides little defence for his account of memes. Its pivotal position in his work appears rather to derive entirely from his citation of Dawkins's original hypothesis. More importantly, Dennett's interpretation of this original hypothesis is often questionable.[11] There are two strands to the source of his major error concerning the relation between memes and consciousness, and both stem from his apparent misunderstanding of a fundamental element of the gene-meme analogy. The resolution of these errors reveals that his explanation of consciousness, as the product of memes, is not as convincing as it might at first appear.

Dennett claims that consciousness developed from simple communication skills when our ancestors learned how to talk to themselves. The advantages that this conferred meant that the "virtuosos" amongst them were selected, and swiftly developed the art of talking to themselves silently. Habits of communication evolved as the cooperative (and therefore successful) members of the community learnt to share the "good tricks" they discovered by this primitive version of thought. "Once our brains have built the entrance and exit pathways for the vehicles of language," says Dennett, "they swiftly become *parasitized* (and I mean that literally, as we shall see) by entities that have evolved to thrive in just such a niche: *memes*."[12] Our brains, equipped with the capacity to communicate with themselves and each other, provide shelter and transmission media for these new replicators: "The haven all memes depend on reaching is the human mind."[13]

Yet, this passage continues, "the 'independent' mind struggling to protect itself from alien and dangerous memes is a myth," for "a human mind is itself an artifact created when memes restructure a human brain in order to make it a better habitat for memes." The essence of Dennett's account, then, is that memes find a haven in the human brain, where the human mind is a combination of their effects. In contrast to genes, which exist within organisms and have their effects primarily on those organisms, memes exist externally and have their effects on the internal structure of human brains.

Confusion Number One: Where Do Memes Come From?

One area of confusion, in this part of Dennett's account of memes, is that he fails to explain where memes might have emerged from, in order to parasitize our ancestors' brains as soon as they had developed language. Did they have had some sort of independent existence before their arrival in the haven of the human brain? Even today the same question arises: if the human mind really is the creation of memes, which are formative constituents of it in the same way that genes are formative constituents of the human body, then where could such replicators originate? (This assertion, as noted, forms the tacit basis for Dawkins's purported meme-virus distinction; if false, then it will also negate that distinction.)

Dennett does make one attempt to explain the origin of memes, but unfortunately it is not very illuminating. His suggestion, that memes "depend at least indirectly on one or more of their vehicles' spending at least a brief, pupal stage in a remarkable sort of meme nest: a human mind",[14] is rather puzzling in view of the examples of meme vehicles which he offers a couple of pages previously: "pictures, books, sayings . . . Tools and buildings and other inventions".[15] In what sense might any of these examples be capable of "spending . . . a brief, pupal stage in . . . a human mind"? This confusion about the nature of meme vehicles will emerge as a key weakness in Dennett's account.

Vocabulary

It will be useful to begin by clarifying a couple of important definitions. *Phenotypic effects* result from a combination of genes and their environment. To oversimplify: if someone has a gene "for" blue eyes then, given the right environmental input (nourishment, etc.), the phenotypic effect of that gene would be his blue eyes.

The roughly synonymous terms *survival machine* and *vehicle* were introduced by Dawkins to refer to any protection and propagation system for genes which houses them, is produced when they band together and whose attributes are determined by them; they are also the means via which their constituent genes can make copies of themselves. We, for example, are characterized as the survival machines or vehicles for our genes. More recently, some people have begun to use a third term, *interactor*, for the same concept.

Confusion Number Two: Vehicles Versus Phenotypic Effects

In Dennett's version of the gene-meme analogy, he notes that genes are "carried by gene-vehicles (organisms) in which they tend to produce characteristic effects ('phenotypic effects') by which their fates are, in the long run, determined". With this picture he compares memes: they are "carried by meme-vehicles, namely pictures, books, sayings . . . Tools and buildings and other inventions are also meme vehicles. A wagon with spoked wheels carries . . . the brilliant idea of a wagon with spoked wheels from mind to mind. . . . The fate of memes . . . depends on the selective forces that act directly on the physical vehicles that embody them."[16]

From this passage it is clear that he distinguishes for genes, but not for memes, between their vehicles and their phenotypic effects. In the case of genes, he separates their vehicles from their *effects* on those vehicles, and states that genes are selected by virtue of their effects. In the case of memes, however, he conflates the two and claims that memes are selected by virtue of the *vehicles themselves.*

He is not normally so pessimistic about the gene-meme analogy, and in this case it seems to be due to his lack of clarity about the nature of replicators' vehicles. In particular, he appears to have been confused by a too literal interpretation of the "vehicle" metaphor – and indeed this is the reason why many memeticists now prefer the term "interactor". For Dawkins a "vehicle" (or interactor) is something that houses and protects replicators, enabling them to make further copies of themselves. Genes build interactors as survival machines: they are protection systems which enhance their chances of surviving for long enough to replicate themselves, and indeed incorporate the mechanisms (sexual organs, etc.) for that replication. For Dawkins, then, interactors do house replicators, but the prime purpose for this is so that they can carry information from one generation to the next.

Dennett, however, appears to be using the term "vehicle" in its more conventional sense (i.e., in the sense in which a car or bus is a vehicle), simply to indicate anything that carries genes or memes around. Because he appears to ignore the fact that interactors are *built by* the replicators that they carry, it is easy for him to forget that the purpose of these vehicles is to make further *copies* of the replicators – not simply to carry them around for as long as they happen to survive.

It is worth stressing this point. Recall that interactors are the means by which replicators make copies of themselves, whereas phenotypic effects are the detailed ways in which those vehicles are "tweaked". This means that phenotypic effects are only *produced* by replicators: my brown eyes are the product of my genes. In contrast, interactors are *both* produced by *and* the source of replicators: the interactor of which my brown eyes are part (me) is both the product of my genes, and the source of half of my children's genes.

When it comes to memes, however, Dennett appears to regard some things as meme interactors, which are actually their phenotypic effects – and it is easy to see how this confusion might lead to the claim that artefacts can be the source of memes. Mandolins, for example, are the phenotypic effects of the mandolin meme – but as soon as you call them meme interactors, instead, then the implication is that they are also the *source* of that meme. At this stage it is a short step to the view of consciousness as being shaped by the memes that spring from such purported sources. The reality is that, whilst the mandolin meme will be *selected* via the success or failure of mandolins, it cannot be *replicated* by them alone.

Confusion Number Three: Representation Versus That Which Is Represented

Here it is important to recall that memes must be *generally applicable* concepts: in mathematics, for example, when I acquire a new meme it endows me with the ability to solve any example of a given type of problem; I do not merely acquire the memory of how to solve those already encountered. When it comes to human artefacts, this distinction is crucial. It is the design or blueprint for an artefact which contains generally applicable information about the construction of that *type* of object. The artefact itself does not contain any such information – and this is as true of the artefacts that Dennett cites as examples of meme vehicles (tools and buildings), as of any other. That crucial information is represented in the blueprint or design from which artefacts result.

Now, it is the job of interactors both to protect and to provide the copying mechanism for the information contained in the replicators that produce them. Clearly no interactor could fulfil this role unless it contained some *representation* of the relevant information – and the crux of my argument against Dennett will be that artefacts do not do this: a bridge contains no representation of its design, nor a mandolin of the concept of that instrument. The confusion of meme vehicles with meme effects is essentially a failure to distinguish between a representation and the thing that it represents.

Dennett is not alone, it should be noted, in believing that artefacts can incorporate memes. Susan Blackmore characterizes memes as "instructions embedded in human brains, or in artefacts such as books, pictures, bridges or steam trains."[17] Rosaria Conte shares my view of a meme as "a symbolic representation of any state of affairs",[18] but then goes on to say that an artefact can incorporate a meme, even if its content is not easy to decode.

So why do I disagree? To find the answer, we need to begin by taking a closer look at what it means to say that a certain characteristic (eye colour, neck length, etc.) is the result of a gene (or gene complex) *for* that feature. One of the things that it means is that there is (or has been in the past) *variation* amongst the genes that control this feature. As Dawkins puts it, "Unless natural selection has genetic variation to act upon, it cannot give rise to evolutionary change. It follows that where you find Darwinian adaptation there must have been genetic variation in the character concerned."[19] This means, further, that the content of any given gene for that characteristic will be partly defined by its differences from the alternative genes for the same feature – its alleles. A significant fact about the gene for blue eyes, for instance, is that it is an alternative to the gene for brown eyes. The same thing will be true of any replicator: its content will be partly dependent on the *differences* between itself and its alleles.

Now, if you show me a token of a spoked wheel and ask me to "build another one of these", then – unless I am already familiar with other wheels – I shall have no reference against which to judge which of the artefact's features will be essential elements of the copy. In order to comply with your wishes, I have to know that this artefact belongs to the general category "wheel", with all that this implies about its function, and so on, and in addition I must be aware of the significant ways in which this type of wheel differs from others. In other words, it is not the wheel but my mind that carries the salient information. In order for me to be able to replicate the meme for an artefact, I need already to possess a wealth of information about that type of artefact – information that will enable me to extract a generalized concept from this particular item.

The difference between this situation and the situation in which you show me a blueprint for the wheel is that the blueprint *tells me* what are the essential elements of the concept. The blueprint is a representation of a spoked wheel, carrying generalized information about that type of artefact; the wheel itself is merely a particular token effect of the information that is represented in the blueprint. An artefact like a wheel cannot be a meme interactor, for it is not able to facilitate the replication of a meme. In order for the meme to be copied, not only does an artefact need to come into contact with a human mind: the mind also needs to create afresh the relevant information, by extracting salient general features from this particular artefact – and it cannot do this without the help of existing knowledge which *it*, not the artefact, brings to the situation. (In this sense, then, the meme is not being *replicated* at all; see Chapter 9 for expansion of this point.) Representations, and not artefacts, realize generalized information – and artefacts can persist long after the information that gave rise to them has disappeared.

Unfortunately for Dennett, the result of his labelling memes' effects as their vehicles is that he does regard things like mandolins as the sources of the memes for them – and once the *cause* of memes has been placed in the external world like this, it is an easy step to the view of consciousness as their internal *effect*.

To recapitulate: Dennett confuses memes' effects with their interactors. In reality, *selection* acts on replicators' phenotypic effects, but *replication* happens via their interactors (in many cases the two overlap, but not always). Dennett refers to phenotypic effects, which are actually the product of memes, in terms that imply that they are also the source of those memes: their interactors. An artefact cannot (usually) fulfil the role that he assigns to it – that of an interactor – since it contains no representation of the information to be copied. In particular, it carries no information about the differences between itself and other artefacts within its general category. Consequently, Dennett is mistaken in thinking that artefacts are the sources of memes rather than their effects. Conversely, therefore, he must be equally mistaken in thinking that the mind is the effect of memes rather than their source.

Dawkins and Dennett

Earlier in the chapter, Dawkins's virus–"good" meme distinction was rejected. The reasoning behind this rejection has become clearer in the discussion of Dennett, whose ambitious account of memes has been criticized as problematic and undefended. Artefacts cannot be meme vehicles – but without an independent external location, memes are *dependent* on minds, and cannot be responsible for building minds in the way that genes build bodies. Of course someone might argue that memes were able to build my mind because their residence in other minds gives them an existence independent of me, but this just takes the question backwards in time: what was the independent external location from which memes colonized our ancestors' minds in the first place? Without entering the realms of science fiction, it seems that denying the possibility of artefacts as meme vehicles effectively rules out the possibility that memes have an external existence, independent of human minds.

Yet if they do not, then we are left with the fact that memes cannot be the formative constituents of the human mind in the way that genes are the formative constituents of our bodies. This, in turn, rules out the coherence of the concept of mental viruses, which are left with no "normal" formative process to parasitize. Even though it seems obvious that the mind cannot develop its full potential without the stimulus of culture, this is a far cry from the claim that memes *create* the mind.

Memes and the Mind

An alternative view is that we are born with the potential for a mind (as part of the brain's neonatal structure: i.e., the product of our genotype), and this is developed via interaction with our environment, a significant element of which is cultural. Replicators must always interact with an environment in order to produce their phenotypic effects, but what I am suggesting is a sea change in our view of memes' role in the creation of the human mind. Rather than as replicators which are its formative constituents, I regard memes as part of the *environment* that contributes to the formation of the mind. To caricature the situation: although it is fair to say that "genes plus environment equals body", the true picture is not so much "memes plus environment equals mind", as "body plus environment equals mind", where memes are part of the latter environment.

This may seem to be at odds with the main force of the meme hypothesis: that there is truly a new form of evolution going on in culture. Surely if I admit that the mind is ultimately the product of our genotype, and add that memes are ultimately dependent on the mind, then I am implicitly admitting that memetic evolution is ultimately dependent on genetic evolution. Fortunately, this is not the case. Even though DNA could not have evolved without the prior existence of carbon, and in this sense it was the existence of carbon which set the scene for the emergence of the organic world, nonetheless it is still true to say that DNA is the source of the biosphere, and that if we want to find out about biology then this is the level at which we need to investigate. Similarly, then, even though the human mind could not have evolved without the prior existence of DNA, and in this sense it was the existence of DNA which set the scene for the emergence of culture, nonetheless it is still true to say that the mind is the source of culture, and that if we want to find out about culture then this is the level at which we need to investigate. It is always interesting to study the relation between different levels of enquiry (chemistry, biology, psychology, etc.), but this does not mean that investigations at one level can be reduced to investigation at another.

The question of the relation between memes and the mind is, as mentioned previously, one on whose answer memeticists are sharply divided, and there is no implication that at this stage I have done anything other than briefly state my own position. The justification for that position is developed throughout the rest of the book.

Where Are Memes?

The failure of Dennett's hypothesis is in a way a shame for meme theory, since his view would have given memetics a formidable corollary: the explanation of the emergence of consciousness. Moreover, having rejected the claim that memes are exclusively external representations, with effects on the internal structures of the human brain, the theory of cultural evolution has been left with no place for its units of selection – or even for their phenotypic effects. In fact the situation looks rather confused: artefacts are denied vehicle status, but other external objects (such as blueprints) seem to have been accepted. It is time to address the question of memes' location.

On the surface, it seems that things are more straightforward for genes. They are found within survival machines, on which their phenotypic effects are exercised to their replicative (dis)advantage. In reality, however, the situation in biology is rather more complicated than the previous sentence would indicate, and the remainder of this chapter explores the relationship between genes and their effects. The implications of that discussion, for memes, will draw together various strands from the preceding chapters, concluding that in terms of memetic location there is no significant distinction between copies that are found internally, in the human mind, and those in external stores of information like libraries and the Internet – but that in order for a meme to be available to selection, *active* copies of it must exist.

The Extended Phenotype: Genes

Dawkins is perhaps best known for his theory of the selfish gene, but his next book, *The Extended Phenotype*,[20] presented a hypothesis that took an equally radical perspective on biological evolution. Like the selfish gene, the extended phenotype is a theory that can usefully be applied to memes as well as to genes. A careful comparison of its implications for culture and biology will elucidate both memes' location and the relations between memes, the mind and culture.

"The doctrine of the extended phenotype is that the phenotypic effect of a gene (genetic replicator) is best seen as an effect upon the world at large, and only incidentally upon the individual organism – or any other vehicle – in which it happens to sit."[21] Genes' phenotypic effects can "include functionally important consequences of gene differences, outside the bodies in which the genes sit".[22] On this view, the concept of a "phenotype" should be extended to include not only those effects which genes exercise upon their own survival machines, but also those which are exercised on the world at large. Beavers' genes, for example, build both beavers and dams, and spiders' genes build both spiders and webs: both sorts of effect are likely to influence the success of the genes that produce them, regardless of their location.

More than this, genes may have effects not only on some inanimate part of the organism's environment, but even on another organism. One example is the effect that a parasite has on its host: Dawkins considers reports that snails infected by trematode parasites, or "flukes", have thicker shells than their uninfected counterparts. "From the point of view of snail genetics, this aspect of shell variation is under 'environmental' control – the fluke is part of the environment of the snail – but from the point of view of fluke genetics it might well be under genetic control: it might, indeed, be an evolved adaptation of the fluke."[23] This is because the optimum shell thickness is not likely to be the same for flukes as for snails: whereas snails value reproduction as well as survival, flukes merely want the snails to survive; therefore a fluke will benefit from a very thick snail shell, even if the resources that go into building it are taken from those needed to maintain the snail's reproductive potential.

So the phenotypic effects of genes may occur in the genes' "own" organism, in the environment or even in a different organism. Via all of these media, the genes facilitate their own survival and replication. The following view of genes and organisms results: "The integrated multicellular organism is a phenomenon which has emerged as a result of natural selection on primitively independent selfish replicators. It has paid replicators to behave gregariously. The phenotypic power by which they ensure their survival is in principle extended and unbounded. In practice the organism has arisen as a partially bounded local concentration, a shared knot of replicator power."[24]

Organisms, then, are the shared effects of particularly intimate groupings of genes, the results of natural selection's "preference" for gregarious replicators. Organisms merely happen, because of this intimate grouping, to provide partial boundaries for the extent of genes' phenotypic effects. In answer to the question why the groupings should have occurred at all, Dawkins replies that, since successful replicators are the ones whose effects depend on the presence of other replicators that also happen to be common (and therefore successful), the world tends to become populated by mutually compatible sets of successful replicators. In principle, though, there are no such boundaries: the whole of the natural world is the product of interactions between the phenotypic effects of its constituent genes.

The Extended Phenotype: Dennett

Before moving on to investigate the application of this theory to memetics, I want to take another brief look at Daniel Dennett's vision of the mind as a meme complex. Although I have been critical of this hypothesis, it should be noted that Dennett enlists the extended phenotype in its support. Since I, too, think that Dawkins's second major claim about evolution has much to add to his first, perhaps it is time to review my earlier opinion of Dennett's arguments.

Recall Dennett's version of meme theory: he says that although we like to think of ourselves as creating, manipulating and controlling our ideas, "even if this is our ideal, we know that it is seldom if ever the reality, even with the most masterful and creative minds."[25] We do not manipulate or control our memes: rather we *are* our memes; they are the creators of our consciousness. In this sense there is no battle between "us" and our "invading memes" any more than there is between our bodies and our genes.

In order to argue that memes have created us, not vice versa, Dennett makes reference to the concept of the extended phenotype. Just as that thesis tells us that webs and dams are a part of their creators' phenotypes, so Dennett believes it to imply that humans' illusory sense of *self* is "a biological product",[26] spun by the brain as automatically as a spider's web is spun by the spider. The human phenotype – which Dennett defines as "the individual organism considered as a functional whole"[27] – does not comprise the body alone, but can be extended to include a "vast protective network of memes".[28]

The Mind: Product of Genes or Memes?

The problem with this claim stems from Dennett's incorrect definition of a phenotype as "the individual organism considered as a functional whole". In reality, a phenotype is "the bodily manifestation of a gene, the effect that a gene, in comparison with its alleles, has on the body, via development".[29] Although we do sometimes refer to an organism as a phenotype, this is loose terminology. The key to the definition of a phenotype is that it is *caused by genes*, as a result of their interactions with the environment. Yet if the mind is a meme complex (Dennett's first claim) then memes are its formative constituents, and it *cannot* be part of the human phenotype (Dennett's second claim), for if it were then genes would be its formative constituents.

Whilst it is true that there will always be two factors in ontogeny – the replicators and the environment – only one of them can exert what Elliott Sober[30] calls the "positive main effect". Which of the factors deserves that title becomes apparent if we imagine alternately holding one constant and varying the other: the positive main effect is the factor on whose variation or constancy the developmental outcome depends. This rather abstract concept becomes clearer when seen in practice.

According to Dennett's first claim, the mind is a meme complex, so memes must be the positive main effect – the most influential factor – in its creation. Thus if it were possible to subject two individuals with differing genotypes to identical memetic exposure, then they would develop roughly identical minds; conversely, if the genotype were identical but the memetic input varied, then their minds would greatly differ. The development of the mind, according to this claim, depends primarily on the memes to which an individual is exposed.

According to Dennett's second claim, however, the mind is a part of the human phenotype, and genes must therefore be the positive main effect – the most influential factor – in its creation. Thus if it were possible to subject two individuals with differing genotypes to identical memetic exposure, then they would develop different minds, whereas if the genotype were identical but the memetic input varied then they would develop roughly identical minds. The development of the mind, according to this claim, depends primarily on the genes that an individual possesses.

The consequences of the two claims are obviously incompatible. Identical twins, separated at birth and exposed to very different memes, would end up with roughly similar minds according to one of Dennett's claims, but very different minds according to the other. He cannot have it both ways.

The Extended Phenotype: Memes

Dawkins's theory of the extended phenotype can, nonetheless, usefully be applied to memes. The essential feature of his theory is that there are in principle no restrictions on the reach of a replicator's phenotypic effects. Although genes are to be found in the organisms that they build, their effects are not limited to those organisms but may also be found in the environment and in other organisms. So we have a three-component picture: genetic information is stored in DNA; it controls the construction of a protective vehicle; it may produce effects both inside and outside that vehicle. The doctrine of the extended phenotype (the third component of the picture) seems surprising because it involves genes having effects on things that they have not built.

This distinction is not relevant to memes, though, since they do not have the second component of the picture: they do not construct survival machines. They do, though, share the genetic distinction between information storage and its effects – and another way to express the theory of genes' extended phenotype is to say that there is in principle no limit to the places where genes' effects may be found, regardless of the fact that the genetic information itself is "stored" internally. So the question posed for memes, by the theory of the extended phenotype, is where we might find their effects: in the mind of their possessor, in the environment or in other minds?

As a starting point, it seems obvious that the acquisition of novel concepts or skills will affect our ways of thought and behaviour. In other words, it is clear that novel memes do have internal effects on the minds of the people who possess them. Similarly, the phenotypic effects of memes on the environment are plain to see: bridges, forms of poetry, methods of central heating, models of the double helix, and so on. What of the purported effects on *other* people's minds (i.e., not on the meme's "possessor")?

Consider the meme for arguments based on reductio ad absurdum: Emma leads Amanda on through a maze of her own opinions, ensuring that she agrees with every step made, until eventually the ridiculous conclusion that results is revealed. Both Amanda and any onlookers may, because the combination of her opinions appears to lead logically to a ridiculous conclusion, come to reject what they judge to be the culprit opinion. In this way, Emma's possession of the reductio meme has the effect that Amanda's actions will protect and propagate some of the rest of Emma's memes, even at the expense of (one or more of) Amanda's own.

This, though, is merely a particular example of the more general phenomenon of manipulation: one person's actions leading to another's unsuspecting cooperation. Such behaviour can frequently be observed, especially in the field of advertising where, for example, if I exploit your desire to look fashionable, and persuade you that my brand of clothing is the coolest, then you will wear clothes with my brand name plastered all over them. At no further cost to me, your action spreads the meme for wearing my brand everywhere you go – and *you* have paid *me* to do the bulk of my advertising.

There are abundant cases of manipulation of one person (or many) by another, with the result that the manipulated spread the memes of the manipulator. The abuse of one's position of authority in one area (e.g., science), in order to promote an opinion about another (e.g., religion), is a further example. Similarly, there is plenty of evidence of memes' phenotypic effects both in their possessors and in the inanimate world: the memetic phenotype may be found in both the internal and the external worlds. So what does this tell us about the location of memes themselves?

The Result

According to the theory of the extended phenotype, it tells us precisely nothing, for it implies that whether memes are realized internally or externally they can have effects in both worlds. This leaves open the possibility that there might be *some* external memetic realizations, as well as some internal copies. Memes, as representations, may be found both within human minds and outside them, in information stores like books and blueprints. If this is the case, then what is the relationship between such external representations and our internal brain structures: are the external realizations merely passive effects of what goes on in our minds, or do they play a more active role in memetic replication?

On the one hand it is clear that if there were *only* external meme stores, then memes could no longer be disseminated. Even if, for example, there are hundreds of copies of a particular theory, stored in libraries all over the world, that theory will have no effect if nobody ever reads it. Similarly, though, the information retained in a human memory may remain inactive for long periods of time. On the other hand, if there were only minds and no external RSs in which information could more permanently be stored, then memetic replication would lose much of its present stability.

The most helpful picture of memetic location may, then, be roughly described as follows: there is no significant distinction to be drawn between the human mind and external information stores such as libraries and the Internet, but in order for a meme to be available to selection, *active* copies of it must exist. If the human mind is not universal, but is developed via interaction with existing culture, then external representations play an essential role in memetic replication. The internal brain structures are, though, the ultimate source of the external representations. Thus a combination of both sorts of meme store has led to a massive capacity for information dissemination and copying stability, which would have been impossible via only one of the storage methods. What matters for both is that the realizations should be of an appropriate kind.

This picture ties in with the view of the capacity to gain and retain attention as the best measure of memetic fitness. If a meme is to be replicated, then it must be able to grab our attention: at times when only passive copies of it persist, it is not able to do this and is therefore not at all fecund. On the other hand, if a meme is to persist then it must be able to retain our attention, and passive copies of it are the most efficient way of ensuring its prolonged existence. This extension of memes' phenotype is also reminiscent of Clark's view that "much of what we commonly identify as our mental capacities may . . . turn out to be properties of the wider extended systems of which human brains are just one (important) part."[31]

What, then, can we conclude about memes' location? Both memes and their effects are to be found inside the human mind as well as outside it, but this is not to say that the two phenomena are indistinguishable: memes are realized in systems of representation, and their effects are not. Cultural evolution depends on the distinction between the two, just as natural selection is ultimately dependent on the existence of discrete biological replicators. In both culture and biology this leads us to ask from where the replicators could have emerged – how evolution could have started in the first place – and I turn to this question in Chapter 10. Before that, however, I want to continue with the task of setting my own views within their cultural context, by examining the work of some other well-known memeticists.

9

The Meme's Eye View

One of the most celebrated commentaries on the meme hypothesis has been provided by the psychologist Susan Blackmore in her 1999 book *The Meme Machine*. Blackmore, like Dawkins and Dennett, accepts that the distinction between virus and replicator is as valid in culture as in biology. Like Dennett, too, she believes that the mind is a meme complex. It is impossible to untangle this mistake from various other strands of Blackmore's thesis – just as it remains inextricably linked with Dennett's confused perception of vehicles and phenotypes, and with Dawkins's erroneous overextension of the virus-replicator distinction – and thus I shall challenge the elements of Blackmore's thought which lead her to share Dennett's view. In particular she focuses on the issue of imitation, to which she assigns enormous significance. Other commentators like Dan Sperber, Robert Boyd and Peter J. Richerson have vehemently disagreed with her analysis, and this chapter also considers their views in the light of what Blackmore has to say.

Copy-the-Product Versus Copy-the-Instructions

I return first to the thorny issue of memes and their effects, which Blackmore acknowledges as an area of confusion when applied to culture. The confusion arises, she says, because of the desire to make an inappropriately close analogy between genes and memes. In the case of memes, she believes that it may be better to abandon altogether the attempt to distinguish sharply between replicators and their effects. Rather, she introduces the concepts of "copy-the-instructions" and "copy-the-product", as a more useful distinction to draw between types of memetic process.

Sometimes, she says, we acquire new information by working backwards from what someone else has produced: for instance, we might watch someone making soup, and later do the same ourselves. In this case we have copied-the-product. At other times, though, we acquire our information more directly, as when we follow a written recipe for making soup. Here we have copied-the-instructions. In cases of copy-the-product, variations will persist if introduced by the individual who is being copied: I shall copy any mistakes made by the soup maker, just as I shall copy her actions when she follows the recipe faithfully. When *instructions* are being followed, however, it will not matter if I see someone alter the recipe when she is making soup; when it comes to my turn to make the soup I shall still follow what is written down in the original, and her alteration will therefore not persist.

This, says Blackmore, is a useful way of looking at what goes on when memes are copied, whereas it is both unrealistic and unhelpful to raise the question which elements of the cultural world are replicators and which their effects. In conclusion, then, she rejects the concept of phenotypic effect as it applies to memes, saying that she "cannot give it a clear and unambiguous meaning".[1]

Nonetheless, it *is* possible to give it such a meaning, and consequently to retain the concept of phenotypic effects within the meme hypothesis. From this perspective we can sketch an alternative view of the distinction that Blackmore characterizes as "copy-the-product vs. copy-the-instructions", and the strength of this alternative interpretation will be demonstrated by both its obvious utility and the coherence of the explanations that it generates.

Not Copying the Product

The previous chapter asked whether artefacts can be meme vehicles, and criticized Dennett's opinion that they can fulfil this role. It suggested that there may in fact be no real replication going on when someone extracts information from an artefact: rather, there is a sense in which he is recreating the information for himself. It is now time to explore this point further, as it is closely related to cases which Blackmore describes as "copy-the-product".

Broadly speaking, there are three ways in which I can create an artefact, whether soup or a mandolin: I can work from an idea that I have invented myself; I can work from instructions that someone else has prepared; or I can copy a product that someone else has created. The first two processes are relatively unproblematic, but difficulties arise in the third case because I have been given no instructions. Adequately compiled instructions – in the form of engineering plans, a soup recipe, or whatever – contain two types of information: instructions about how to make the end product *and* information about what its essential features are. An artefact, on the other hand, contains no information about which of its features are accidental or aesthetic, and which are essential to its function. If I wanted to make a copy of a wooden spoked wheel, for example, then no matter how closely I examined it, there would be no means of my telling whether it was important that the wood's grain ran along the length of the spokes rather than across them – or whether that was just the work of a particularly meticulous craftsman. There is, then, information in a blueprint which is just not present in the object that it describes.

Now you might, of course, already possess some of that missing information yourself. As an experienced cook, it might be immediately apparent to you which are the unique features of that particular soup recipe; as a trained engineer, the importance of grain direction for strength may seem to you blindingly obvious. If you had enough of the missing information, then you could probably copy the product: this is what happens all the time in manufacturing industry, where competitors' products are routinely analysed and dissected for comparison and inspiration. The point, however, is that in these cases – which Dennett would see as artefacts fulfilling the role of meme vehicles, and Blackmore would describe as copy-the-product – the relevant information has been brought by *you*, rather than gleaned from the artefact. It is not possible to generate, from an end product, information about which of its features are relevant or significant: if you want to copy that product, then you need either to have access to its plans, or to *bring* to the situation the information that you would otherwise have gathered from the plans. If you do the latter then you have obviously not *copied* that information, for you already *had* it.

Even if you made an exact copy of the product, correct grain direction and all – simply by mindlessly reproducing everything that you observe, without knowing its significance – still you would not have gained a copy of the relevant information. You may, in so doing, produce a set of instructions for repeating the product-copying process, but this would not be a copy of the original information: again, it would be information that *you* had originated, in the process of copying the product. It would include your own inferences about which features are significant, but this is not the same as a copy of the *original* information about significance. The inferences that different people would make in the same circumstances may be extremely varied, since each of them would bring to the situation his own range of experience and level of deductive skills.

Thus, it is arguable that if I learn to make soup by watching someone else (i.e., copy-the-product) then I am not truly copying the information *from her* at all, but rather am re-creating it for myself. Unlike when I read the recipe, which contains a representation of the relevant information, my own soup making will be guided by my *inferences* from what she is doing to the instructions that she was following – not by the instructions themselves.

Memes and Their Effects

The key to the problem is that, just as a spoked wheel is the phenotypic effect of a meme for that concept, so the actions involved in following a recipe, together with the soup itself, are the phenotypic effects of the information contained in that recipe. "Copying-the-instructions" is, on this view, a truly replicative process, via which I acquire a meme (however briefly: it may be that I only ever make this soup once) by coming into contact with an existing copy of it. What Blackmore calls "copying-the-product", on the other hand, is not really a copying process at all, since here I acquire information by a process of inference from phenotypic effects to memetic content, using information that I already possess: any new representations that I form as a result of this process do not come from anyone else.

Blackmore's distinction between copying-the-product and copying-the-instructions therefore presents no challenge at all to a cultural distinction between replicator and effect. Rather, it is an alternative way of viewing the same phenomena, and one which (by calling both processes a type of copying) manages to obscure what is really going on. The distinction between a meme and its effects remains valid and useful.

Memetic "Drive"

Fortunately for Blackmore, her distinction between copying-the-product and copying-the-instructions is really a corollary to her version of memetics. More fundamental to it are the concepts of imitation and memetic "drive". She assigns great significance to imitation as the copying mechanism on which cultural evolution depends; indeed she would *restrict* memetic replication to imitation, saying that other forms of learning are not adequate to the task. The next section will look in more detail at this claim; here I focus on the related concept of memetic drive, a process via which memes are said to change the environment for genes.

Blackmore begins by pointing out (uncontroversially) the genetic advantage to our ancestors of being able to imitate others' useful behaviour. As a result, she says, those who were best at imitating – "meme fountains"[2] – would have had a particular advantage, and thus their genes would have spread. In these circumstances, in addition, others would tend to "copy meme fountains and their popular memes", and hence the meme fountains will acquire "improved power and status". They will therefore succeed both genetically and memetically – and more than this, "If there are genes for imitating the best imitators, these genes will also spread in the gene pool."

As the tendency to imitate proliferates, and people become "better at imitating the successful memes", so culture will expand and memetic evolution will begin to result from competition between varieties of cultural traits. But then *genetic* survival comes to depend on the ability to discriminate between genetically useful and genetically damaging *memes.* So it turns out that the successful memes "change the environment in which genes are selected. In this way, memes force genes to create a brain that is capable of selecting from the currently successful memes."

Memetic drive is thus a phenomenon whereby "successful memes spread. They then change the environment in which genes are selected. The consequence is a brain that is better designed for spreading those particular memes."[3] The brain becomes, in other words, rather like an immune system "for recognizing which memes are useful and which not".[4]

Meme Fountains

A minor point here is that there are problems with the concept of meme fountains. In particular, it is implausible that there should be "genes for imitating the best imitators". If imitation is, as seems likely, an innate human characteristic, then clearly there must at some stage have been variation amongst the genes that controlled *how* we imitated, how *well* we imitated and probably also *what* we chose to imitate – but it is not obvious how our genes could control *whom* we chose imitate.

Indeed, this idea merits further exploration, for it is not even clear exactly what it would mean to "imitate the best imitators". The trouble is that the "best imitators" will not necessarily be a fixed group of people, since different cultural trends will favour the innate abilities of different individuals: technological developments will be more easily picked up by one sort of person, intellectual or aesthetic novelties by another. It is unlikely in the extreme that there could be genes for imitating people who are the best technological imitators, other genes for imitating the best intellectual imitators, and still others for imitating people who most quickly pick up new musical ideas – and that these separate groups of genes could then be selected swiftly enough to keep up with the relevant cultural developments. All of these changes – both the developments within a particular cultural area and a society's general shift of focus to a different cultural area – will happen much too quickly to be picked up at the genetic level.

It may be that the idea of meme fountains seems to makes sense because at one level it almost expresses a truism. That is, people will (because they want to gain status) tend to copy the most popular behaviours – and the people who *already* engage in those behaviours must be those who were able to pick them up most quickly in the first place. It certainly is plausible, for example, that there should be genes for *being* the best imitators of technological or musical novelties – and that these might give their possessors a social advantage at times when cultural evolution is favouring those areas. At any given time, then, there will be a certain group of people who (because they find it very easy to pick up the current cultural novelties) are those whose behaviour is most often copied: in any chain of imitation it is of course the case that some people are nearer the beginning than the end, and it is plausible that genetic make-up will help to determine one's place in the chain. This group, though, will be a shifting rather than a fixed collection of people, and the genetically favoured trait will be the tendency to imitate the currently popular *behaviour* (which by definition has already been adopted by a group of people, otherwise it would not be called "popular"), rather than to imitate the particular people in whom it can be observed.

Religion as an Example of Memetic Drive

Despite this niggle, the idea of meme fountains is arguably not crucial to the idea of memetic drive, and I return now to the core of Blackmore's hypothesis. The concept of memetic drive plays a key role in her explanations of a variety of phenomena, including the development of the human brain and the origins of language and religion. Here I consider her analysis of religion, as an example of the use to which she puts this concept, and as a means of exploring its validity.

Blackmore claims that "when we look at religions from a meme's eye view we can understand why they have been so successful."[5] Her idea is that religions are memes, and that as a result of the power and status that accompanied their religious behaviour, the most religious people may have been the most successful in finding mates. In this environment memetic drive would ensure that *genes* for religious behaviour – especially genes for "the kind of religious behaviour best suited to" the religion of the time[6] – would also flourish.

How would this have worked? In other words, how exactly could religious memes have driven our genetic evolution? Blackmore's suggestion is that a spectacular coincidence, such as an apparently answered rain dance or supposedly miraculous cure, would bring enhanced power and status to the individuals involved. Other people would then copy the apparently successful behaviour (with some variations), in the hope of gaining some of that power and status for themselves. The people whose variations were the most flamboyant, or coincided the most often with the desired outcome, would not only see their *memes* succeeding as a result (i.e., their versions would be the ones that most other people chose to copy), but would also attain a *genetic* (i.e., sexual) advantage, as a result of their increased power and status. Consequently, any genes that were involved in the control of those particular versions – for example, genes that gave individuals an advantage in flamboyant dancing – would spread throughout that culture. The cumulative effect of these processes would be a brain genetically tailored to the acquisition and imitation of religious memes.

Problems for Memetic Drive

I am unconvinced by Blackmore's exploration of memetic drive, for two key reasons. First, the speed at which memes evolve makes it implausible that specific memetic developments *could* act as selection pressure on genes: memetic changes will usually be far too swift to be picked up at the level of genes. "Depending on population size and the intensity of selection, the specific adaptive demands imposed by the environment must remain unchanged over hundreds or even thousands of generations in order to produce the level of gene replacement necessary to cause a new trait to become a regular feature of a species."[7] This means that a population would have to maintain a roughly invariant form of religion for millennia in order for genetic selection to catch up.

Leading on from this point, and perhaps more significantly, it should be noted that natural selection never *could* catch up, unless there were genetic variation for the relevant traits. Of course memetic variations will sometimes provide their bearers with genetic advantages: think of the meme for rejecting contraception, which will be genetically advantageous so long as there is sufficient food for all resultant children; conversely, the meme for contraception can itself be genetically helpful when used to keep the population within the limits of its resources. This does not mean, however, that there will be any level of *genetic* control over the advantageous memes – and the problem for Blackmore is that, if there is no genetic variation between those who do and those who do not subscribe to the relevant memes, then there is nothing on which natural selection *can* work.

Returning to her example, this means that only *if* there are genes for religious behaviour, and variation amongst them, can those genes be selected. If not, then although the people who subscribe to religious memes may succeed genetically, the genes that come to prevail in the population as a result need have nothing to do with religion. They will simply reflect whatever other genes happen to make up the genotypes of the successful individuals.

When studying any evolutionary process, Blackmore recalls Dennett's urging us "always to ask *cui bono?* or who benefits? and the answer is the replicators"[8] – but it is important to remember that there are *two* aspects to this question, of which the translation "who benefits?" is only one. The *Concise Oxford Dictionary* highlights both elements, in its definition of *cui bono?* as raising the "question of who stood to gain (and therefore was likely to be responsible)".[9] The problem is that the two elements of this definition can sometimes be in conflict: the answer to the question which sort of replicator is likely to be responsible for an evolutionary development does not always follow in a straightforward way from the answer to the question which sort is likely to gain from it. Thus, identifying the beneficiaries of a process does not necessarily identify the controllers. The *genes* of people who reject contraception will in certain circumstances *benefit*, for instance, from the *meme* that is *responsible* for that idea – but this tells us nothing about whether those people also have genes which make them tend to accept that meme. If not, then there will be no link at all between the genes that benefit from the use of contraception, and the meme that is responsible for that idea.

Similarly, the genes of religious people may in certain circumstances benefit from the memes that are responsible for their ideas and practices – but this tells us nothing about whether those people also have genes with any degree of control over their religion. It may well be that this is another example of memetic responsibility for traits which bring benefit to genes that are wholly unrelated to those traits. It is irrelevant that genes for certain religious behaviours *could* benefit from the genetic success of individuals who subscribe to the memes for those behaviours: if such genes do not exist, then they obviously can neither benefit from, nor be responsible for the success of those individuals. Even if they do exist, moreover, the rate of change in the memes that control the same sorts of traits may override any possibility of selection amongst those genes.

In summary, then, Blackmore claims that "the memes that succeed in memetic competition change the environment in which genes are selected, giving an advantage to genes which help a person imitate the currently successful memes – whatever those memes happen to be."[10] I have questioned the link between "the currently successful memes" and the "genes which help a person imitate" them, on two grounds: there may be no such genes; and even if there are, the changes in the relevant memes will usually be much too swift to be picked up at the genetic level. For these reasons, the concept of memetic drive seems to me to be deeply flawed.

Imitation

At the root of Blackmore's concept of memetic drive is her theory that memetic transmission must work via imitation, because "only imitation is capable of sustaining a true evolutionary process." It is time to examine this claim in more detail.

Blackmore maintains the unique significance of imitation for memetic transmission, because she says that it is the only form of social learning in which true replication takes place. There are, of course, many types of social learning, and like other observers Blackmore notes the distinction between merely reproducing behaviour and truly replicating it. In what is known as "stimulus enhancement" or "local enhancement", for example, the attention of one animal is directed by the behaviour of another towards a location or object in the environment, and as a result it then behaves in a similar way to the animal that it observed. A typical instance of this sort of learning is the behaviour of blue tits and other small garden birds in Britain, which peck at the tops of bottles of milk left on doorsteps. It is well documented that they do this because they have seen other birds doing the same thing. Yet there is no true replication of behaviour here, says Blackmore, because although one individual bird ends up with similar behaviour to another, "the behaviour is not copied": rather, an existing behaviour is reproduced in a novel environment. The blue tit could already peck for food, and has simply learned that here is another arena in which that behaviour will be fruitful.

In contrast to such behavioural reproduction, behaviour is truly replicated when, "by observation of another individual performing an act, an animal is able to reproduce the same motor pattern"[11] – a motor pattern which it had not previously produced. In order to do this, adds Blackmore, the animals must be able to *imitate*, and thus "without imitation there is no replicator and no new evolutionary process."[12]

Now, the distinction between the reproduction and replication of behaviours is widely accepted, but the conclusion that Blackmore draws from it is not. It will be helpful, therefore, to look more closely at what is involved in imitation.

What Is "Imitation"?

The first point revealed by a closer inspection of Blackmore's account is that it seems, at times, to confuse imitation with *replication*. Of course there can be no evolution without replication, but Blackmore's claim is that in culture there can be no evolution without *imitation*. Having argued (fairly uncontroversially) that not all forms of learning will count as true replication, she appears simply to assume that imitation is the only form that will. In reality, a distinction between the reproduction and the replication of behaviour tells us nothing about which learning methods will support replication. Many forms of learning or passing on information could facilitate evolution, *if* they involved replication – and Blackmore seems to be in danger of *defining* them out of the picture, with her statement that imitation is the only form of learning that involves true replication. What about the information that we gain from reading or being taught, for instance? Blackmore would say that teaching, reading and writing are just different forms of imitation, but these processes are so much more sophisticated than the imitation that is involved when a baby waves back at his mother, for example, that I am surprised by her insistence that we stick to just one word to cover all sorts of memetic transmission methods.

It may be, as Henry Plotkin has claimed, that in Blackmore's work "the notion of imitation has been expanded beyond the point of meaning."[13] In reality, says Plotkin, "different psychological mechanisms" are at the base of "the imitation of a motor act, the acquisition of a native language, and learning one's culture-specific social constructs."[14] These are not all instances of the same type of process.

In particular, he draws attention[15] to two different sorts of memes. The first category comprises memes that are "informationally narrow in scope": these are short-lived, situation-specific memes, such as the knowledge that a particular restaurant is good, or the lyrics to a currently popular song. Memes of the second sort are made up of "higher order memories and knowledge structures", and are "of much wider scope informationally, and of much greater longevity, with transmission normally restricted to just once in a lifetime". Overarching concepts like *restaurants* and *songs*, on which the situation specific memes are based, would fall into this category. Such higher order memes "are also closely interwoven with others"[16] – think of interdependent concepts like shop and money – and their transmission "is smeared out over a considerable period of time, yet the replication achieved is probably just as accurate as is an imitated motor act". In Plotkin's view, such higher order memes "are not acquired by imitation but by a complex process of construction and integration" – and it seems obvious that this is indeed the case. Once our attention has been drawn to these sorts of memes, it becomes obvious how poverty-stricken is a theory which would restrict memetic replication to imitation. "Nowhere is Occam's Razor more misplaced than in a science of culture."[17]

Does Culture Replicate At All?

Yet if imitation alone is not adequate to the task of memetic replication, then clearly more needs to be said about the processes that are involved. At the other extreme from Blackmore, writers such as Dan Sperber, Robert Boyd and Peter J. Richerson have challenged the idea that cultural information is truly replicated at all. They accede to her distinction between the reproduction and replication of behaviours, but would use it *against* the meme hypothesis, maintaining that the process at work in culture is almost always reproduction rather than replication: "most cultural items are 're-produced' in the sense that they are produced again and again – with, of course, a causal link between all these productions – but are not reproduced in the sense of being copied from one another."[18]

If this is the case, then the game is up for memetics, for whatever one thinks about Blackmore's claim that imitation is the key to cultural replication, it is certain that cultural evolution does depend on *some* sort of replication taking place. Does her emphasis on imitation stand up to the criticisms that Sperber, Boyd and Richerson level at it, or is the opposite true – that there is no such thing as cultural replication?

Dan Sperber

Dan Sperber presents a threefold argument for denying that most cultural information is truly replicated. The first reason why cultural transmission is mistakenly seen as a process of replication is, he says, that memeticists are overattached to the biological analogy, in which replication is the norm and mutation an accident. Sperber believes that in culture, by contrast, the mutation rate is so high that "the very possibility of cumulative effects of selection is open to question."[19]

This challenge was answered in chapter 5: the crucial point is not whether memetic variation rates are high per se, but whether they are too high in relation to the rate of memetic replication – a process which is itself much swifter than its biological counterpart. More than this, it may be that Sperber has himself displayed an overattachment to the biological analogy, by focusing too closely on the genetic variation that arises through mutation. Mutations may indeed be rare in relation to the rate of genetic replication, but the variation that arises through *recombination* is not. Indeed, sexual reproduction results in the recombination of genetic material *every* time that it is replicated, and yet genetic content is still replicated "well enough to undergo effective selection".[20] The lesson is that we cannot tell whether evolution is undermined by a high rate of variation without reference to facts about replication rates and the means by which that variation arises.

The second reason why Sperber denies that cultural information is truly replicated is that he rejects what he sees as a mistaken tendency in the social sciences to idealize away the individual differences amongst cultural representations. He agrees that it is possible to represent, in a prototypical manner, the partly common content of a chain of sufficiently content-similar representations. The problem, he says, is that it then becomes tempting to see all the individual concrete representations as having the same content, with negligible variations – that is, as (imperfect) replicas of each other. Again this is a point that came up in Chapter 5: there, in the course of discussing memetic alleles, I showed that a meme-based perspective can account for all of the phenomena that he identifies.

Sperber's third argument is that representations are not *replicated* in the process of cultural transmission, but *transformed*. He says that another damaging tendency in the social sciences is to see communication as a process of encoding and decoding information, in which the same information is copied from one mind to another – but that in truth what happens when representations are transmitted is that "human brains use all the information they are presented with not to copy or synthesize it, but as more or less relevant evidence with which to construct representations of their own."[21] Information is transformed in the processes of remembering and communicating, not "copied" as such: it is either "inferred" (when implicit) or "comprehended, a process that involves a mix of decoding and inference" – and in either case, "information provided by the stimulus is complemented with information already available in the system."[22] Thus any coincidence between the information in your mind and the information in mine is due not to any intrinsic properties of the information itself but to the ways in which our minds work.

In developing this argument, Sperber goes further than simply presenting arguments against cultural replication. Given the concerns of writers like Plotkin, who asks whether imitation is the only alternative to the reproduction of behaviour, it becomes likely that there will be more than one process by which true cultural replication can take place. If so, then how are we to determine whether any given process of cultural transmission really counts as replication? Sperber suggests a test to which such processes can be submitted, which will determine the answer to this question. The problem, he says, is that most cases of cultural transmission will fail it.

Sperber's Test for Replication

Sperber's test consists in "three minimal conditions for true replication": if B is a replication of A, then B must be "caused by A", must be "similar in relevant respects to A", and must "inherit from A the properties that make it relevantly similar to A".[23] Thus not everything that is passed on from one person to another will count as an instance of replication: infectious laughter, for example, passed along a line of giggling schoolchildren, will not, since it fulfils only the first two conditions and not the third. Balbir's laughter is caused by Sukdev's, and sounds rather like it – but that *similarity* is not caused by Sukdev's laughter. Rather, both children already had their own laughs before this situation arose: when Sukdev's laugh triggers Balbir's, any similarities between their laughs are due simply to both sounds *being* what we would call "a laugh". Although Balbir's laugh is triggered by Sukdev's, there is no causal link between its properties and those of Sukdev's laugh: it does not, in other words, inherit from Sukdev's laugh that which makes them similar.

Dawkins's Test for Replication

Sperber offers his three conditions for replication as an alternative to the test suggested by Richard Dawkins,[24] who says that replication is present whenever an intelligent observer cannot discover the approximate original order of a chain of causally linked items, as illustrated in the example that follows.

If I ask someone to memorize a picture that I have drawn and then to produce a copy of it, and then I ask another to do the same with the picture that the first one produces, and so on down a line of ten people, then there will be one of two possible outcomes. If the original "picture" was a fairly random scribble, then each person in the line would really try to memorize the drawing of the previous participant, although there would probably be a fair degree of difference in the results. It would be possible, if the pictures were lined up in order, to follow the trail of alterations between participants' drawings. If the picture were, instead, something recognisable like a five-pointed star, then each person would simply try to produce her own drawing of a five-pointed star, rather than to reproduce an exact replica of the previous participant's version of that picture (wobbles, inaccuracies and all). If these pictures were lined up in order, then there would be no trail of alterations to follow, because any variations would not be based on the previous person's drawing.

If I then shuffled the eleven pictures and asked someone to try to put them back into the order in which they were produced, it would obviously be easier in the first case than in the second: this, Dawkins would say, is because copying the star does and copying the scribble does not involve a process of true replication. Whereas the people trying to copy my random scribble were attempting to produce a copy of the drawing itself (i.e., bit by bit, with no understanding of its underlying structure), those copying the star were unconsciously attempting to follow an instruction – "draw a five-pointed star" – and that *information* was effectively replicated by each participant. Dawkins argues that the question whether a sequence of products could be accurately reordered by an intelligent observer is therefore a good test of whether the replication of information was involved in the copying process.

Sperber Versus Dawkins

Sperber, however, undermines Dawkins's test with examples of sequences that would pass it, but which clearly do not involve replication. He points out, for instance, that infectious laughter would pass Dawkins's test – in that someone listening to a jumbled-up recording of the children would be unable to reorder their bursts of laughter – but that laughter is obviously not being replicated along the line of children. He insists that, for true replication, his third condition (the inheritance from the previous generation of the relevant properties) is also necessary.

I find Sperber's counterexamples persuasive, and would suggest that his criterion has an added advantage over Dawkins's: it can distinguish between chains of replications that are related to each other in different ways.

It could distinguish, for instance – where Dawkins's could not – between the case in which each participant was asked to copy the previous participant's star, and a case in which each used the same source, being given my original drawing of a scribble rather than the previous participant's copy. In neither case would an intelligent observer be able to reorder the drawings, and therefore Dawkins's test would categorize each as a case of true replication, without revealing that two very different patterns of copying were involved. Sperber's test would pick up this difference, though, because there is such a different sequence of inheritance in each. The sixth participant's drawing, for example, is a copy of the fifth person's in one case (where each was given the previous person's star), and of mine in the other (where each was given my scribble). The participants' drawings have inherited from a different source, in each case, the properties that make them relevantly similar to my original drawing, and the sequence of replication is therefore different.

Sperber's test would also reveal the difference between each of these sequences and a third one in which each of ten people was asked to draw a star (i.e., given the instruction rather than a drawing to copy), and coincidentally all chose to draw one with five points without lifting pen from paper. Here again Sperber's test would show that this sequence involved a different causal chain from the others, whereas Dawkins's test would not be able to distinguish this collection from the others, but would simply reveal that none of the three sequences could be put back in order.

Back to Sperber

Having proposed his test for true replication, Sperber denies that most instances of cultural transmission will pass it. When people communicate, the desired result is, undoubtedly, a similarity in content between the information in the speaker's mind and the resulting information in the listener's mind. The problem, he says, is that even when such similarity is achieved, the listener's information does not derive *from* the incoming information the properties that make them relevantly similar. Any similarities are due, not to any intrinsic properties of the information itself, but rather to a "constructive cognitive process"[25] in the listener's mind.

Sperber's Test: A Problem for Memes?

Clearly Sperber is right to emphasize the fact that information is not always copied directly from its source: we humans are pretty good at attaining information, even from fairly limited resources. In particular he points to the process of language acquisition, in which grammar is not "present to be copied" by infants, but rather is inferred from incoming data using a "genetically determined preparedness".[26] I would go even further than that, adding that we have a genetically determined preparedness to be good at copying and extracting information from limited data of all forms, not just language. It is in the implication of this claim that I would differ from him: couldn't such innate mechanisms be the *means* by which information is replicated, rather than an *alternative* to replication?

Sperber says that "For memetics to be a reasonable research programme, it should be the case that copying, and differential success in causing the multiplication of copies, overwhelmingly plays the major role in shaping all or at least most of the contents of culture."[27] To a certain extent I would agree with this – but only if it is acknowledged that *copying* cannot simplistically be equated with *imitation*. The problem is that I suspect that his arguments *are* aimed largely at a rather narrow conception of copying, perhaps limited to something like "observational learning". He asserts, for example, that "evolved domain-specific psychological dispositions, if there are any, should be at most a relatively minor factor" in any process of true replication: in other words, instructions are not really copied if the similarity of original and subsequent is due more to the ways in which the observer's brain *interprets* what he sees, than to the nature of the original instructions.

Yet how can I disagree with this? Having accepted Sperber's test for replication, of which the last sentence of the previous paragraph is just a particular application, it would at first glance seem contrary for me to differ from him now that its results are inconvenient for memetics. Indeed, more than accepting his test, I support his view that human brains use incoming information "as more or less relevant evidence with which to construct representations of their own". So how can I continue to maintain that human culture depends on a constant process of memetic replication?

My claim is that the information in your brain can derive *from* the information in mine the properties that make them relevantly similar, *as a result of* being copied via a replicative machinery that includes evolved psychological mechanisms. Such mechanisms can be seen as just one element of the meme-transmission process: inbuilt error-correction systems that help us to receive memes with a fair degree of accuracy, hence facilitating memetic evolution. Even if we rely on such systems in order to receive information, this does not mean that the information so received does not really come from another person. We have not invented it ourselves; it has still come from someone else. A helpful analogy may be drawn here with the process of translation between two public languages: when I read a passage written in French, and have to interpret and reconstitute it in order to understand it in English, still the information that ends up in my brain has come from the passage that I was reading: I haven't invented it just because I had to translate in order to receive it.

Thus I can agree with Sperber that replication has not truly taken place unless my information derives from yours the properties that make them relevantly similar; agree too that it is wrong to assume that "in general, the output of a process of transmission is wholly determined by the inputs accepted or chosen by the receiving organism"[28] – and yet disagree that these facts undermine the cultural replication that is necessary for memetic evolution to take place. Of course we should expect that the processes of cultural replication will be complex in nature, for the information being copied is at times incredibly complex. A relatively simple process like imitation cannot be expected to do the job in all cases: you don't *have* to overemphasize imitation in order to be a memeticist.

Robert Boyd and Peter J. Richerson

Nonetheless, more arguments have been ranged against the view that cultural information is really replicated. Where Sperber focuses on the origins of any similarities between the information in people's brains, Robert Boyd and Peter J. Richerson argue that often those similarities will, in any case, be few and far between.

Although sympathetic to an evolutionary view of culture, Boyd and Richerson deny that cultural evolution depends upon selection amongst replicators, claiming that memes cannot explain the evolutionary changes that occur within culture. The reason is that ideas are not simply "copied and transmitted intact from one brain to another".[29] Rather, someone will observe a behaviour in another person, and then induce the information necessary to produce the same behaviour – and the information thus induced need not be the same as that in the originator's brain. As a result, "the replicator model captures only part of cultural evolution", because cultural change is also shaped by "genetic, cultural, or developmental differences among people".[30]

Replication Versus Inference
In practice, this means that different information can often be inferred from the same overt behaviour, depending on our own cultural background. Boyd and Richerson give the example of a parent and child who both use the same pronunciation for a certain phoneme (e.g., the "wh" sound in "what", "where", etc.), which superficially seems to imply that the child has a *copy* of the parental representation of that sound. In fact, in Boyd and Richerson's example, each has a different mental representation of it: the parent has altered his pronunciation since his own childhood (because he has moved to an area where a different pronunciation predominates), whilst the child shares that pronunciation because it is the only one that she has ever heard. Their representations were therefore formed in quite different ways, and an apparently obvious case of replication turns out not to be an instance of reproduced information at all.

Boyd and Richerson point out that this is not an isolated example; that cultural changes are *often* the result of information being induced from other people's behaviour – a process that is dependent on the psychology and background of the inducer, and which can therefore result in a *different* set of information in her brain from that in the brains of the people whose behaviour she originally observed. Boyd and Richerson conclude that the information represented in her brain is not, therefore, a *replication* of that in the brains of the people she copied, because the two sets of representations have different content.

Indeed, this seems to be a fair assessment of the situation in their particular example, which illustrates the fact that the same observations can lead to the formation of different representations. This fact is not enough, however, to demonstrate that the representations thus formed are not replicators (just that they are not, in this case, tokens of the *same* replicator).

It is unproblematically the case that two different replicators can give rise to the same behaviour, or phenotypic effect, and that we need to know the history of the replicators in question before we can discover whether they are truly the same as each other. Observation of their effects is not enough. This is as true for genes as for memes. For instance, Edward and Elizabeth might both have brown eyes, although Edward has blue/brown alleles and Elizabeth has brown/brown alleles (blue alleles being recessive); each has the same phenotypic expression, although they have different replicators. Moreover, even if both had the same replicators, this alone would not tell us that they were related to each other: for that, we should need to learn more about their history. Similarly, the parent and child in Boyd and Richerson's example may display the same pronunciation without sharing the same representational content, and indeed there is a sense in which the child has not so much copied her parent's representation as invented her own – but this does not in itself comprise evidence against the memetic view of culture. The only thing that this example shows is that parent and child do not in this instance share the same replicator – not that there is no such thing as a cultural replicator.

Replication Versus "Averaging"

The point of this parent-child example, though, can be generalized in a way that presents a more serious challenge to memetics. Boyd and Richerson assert that cultural change can be (indeed, is often) shaped by things *other* than the nature of the cultural information itself. They claim that memes are inadequate to explain the evolution of culture, because the changes that accumulate towards that evolution are more often the results of differences amongst people than of the nature of the cultural information itself. Cultural evolution, in other words, is directed by population processes rather than by discrete cultural replicators.

Now, there is a sense in which my own view simply offers a different perspective on the same facts. I would characterize population processes – that is, the differences between people within a population – as part of the memetic environment: they are one element of the selection pressures acting on memes (as discussed in Chapter 6), rather than a problem for the meme hypothesis. Moreover, I would emphasize that not all cultural change depends on observational learning, in which information is inferred from observations of its effects, as in Boyd and Richerson's example. We can also attain information more directly, through teaching and other forms of linguistic communication, in which cases individual idiosyncrasies have less of a mutating effect on the information transmitted.

But Boyd and Richerson, too, would take their claims a stage further, maintaining that memes are not only insufficient to explain cultural evolution; they are also unnecessary. Sticking with the example of linguistic information, they raise the question *how* children will choose which pronunciation to adopt, in a population where there is variation across a particular phoneme (e.g., the "a" in "bath"). Imagine, for instance, the situation of children with Yorkshire parents, who attend a nursery where the staff include people from London, Devon and South Africa. How do these children settle upon an accent of their own? Perhaps they select an adult and copy his particular pronunciation, in which case they have replicated his particular mental "rule". Or perhaps they average out what they hear across the population, and adopt the rule of an *average* pronunciation – in which case, say Boyd and Richerson, there is no replicator, because no mental rule is transferred faithfully from one brain to another. The latter possibility leads them to claim that a key question for those who would defend memetics is, "When a child has the chance to copy the behaviour of several different people, does she choose a single model for a given, discrete cultural attribute? Or, does she average, or in some other way combine, the attributes of alternative models?"[31]

This is an interesting question, but I do not agree that its answer will prove of great significance for the meme hypothesis. As before, Boyd and Richardson have offered a different perspective on the same facts, rather than presenting a devastating challenge to memes. From a memetic perspective, the "averaging" rule (if it exists) is simply one of the influences on *how* memes are transmitted – in this case an influence that usually counts against fidelity by introducing variation on almost every copying occasion. Nevertheless, from this point of view it is simply not true that the same information has not been transmitted: rather, what has been transmitted is a novel *allele* of the *same* replicator. The information in this case is a representation of a phoneme, and phonemes may be articulated in different ways within a population whilst still being regarded as identical by native speakers. This means that any of the phoneme's variants (known as allophones) may be used without affecting the meaning of what is said. In the example outlined, a mental rule could be replicated just as well via the "averaging" as by the more straightforward method. The only difference would be that "averaging" transmits a different allophone of the phoneme being copied, and in memetic terms this would be described as the replication of a different allele of the same meme.

Furthermore, the "averaging" rule could only work in a way that is actually *dependent* on replication: how else could the children "average" the incoming information, if it were not first copied into their brains? Indeed, "averaging" is a process that compares very closely to the recombination of particulate genes. The children, on this view, receive copies of a variety of alleles of a certain meme, and by a process of recombination arrive at their own chosen version. Indeed, it might even be said that they acquire all of the alleles, only one of which becomes dominant and hence is able to exercise a phenotypic effect on their pronunciation. On this view the others are recessive in those individuals: they are aware of their existence as mutually replaceable alternatives to their own ways of talking, but those alternatives have no effect on their vocal behaviour.

Conclusions

Boyd and Richerson have claimed that replicators cannot account for evolutionary changes in culture, because cultural information is not replicated as such but instead is often inferred from observations of other people's behaviour – which leads in many cases to the creation of different information in the brains of those involved. Indeed, they say that at least some of the ways in which cultural information is transmitted (e.g., "averaging" the variants in the surrounding population) actually prohibit the faithful replication of discrete cultural attributes. Memes cannot explain the changes that occur in these cases, because they are the results of differences amongst people, rather than of the nature of the cultural information itself.

There is, as mentioned, a sense in which Boyd and Richerson are merely looking from a different perspective at the same facts as meme theory purports to explain. More than this, their emphasis on population thinking could provide a useful foil to some other theorists' tendency to overstress memes' autonomy. Nevertheless, their examples and arguments do not present a valid challenge to memetics, which actually predicts that there will be variation amongst versions (alleles) of particular pieces of cultural information (memes). Nor is any challenge to memetics presented by the facts that cultural information can mutate during the transmission process, and that different mental representations can lead to the same overt behaviour. The differences amongst the human population provide an important influence on the accuracy and speed of memetic transmission, but Boyd and Richerson have overstressed the significance for memetics of population pressures on cultural change.

Imitation: A Recap

Blackmore has drawn attention to the distinction between the reproduction of behaviour and genuine imitation. Her claim is that imitation forms the basis for all memetic replication, but writers like Plotkin have challenged her belief that imitation is the *only* alternative to behavioural reproduction. Some forms of information transmission are so much more complex than what we normally intend by "imitation" that the question arises what *is* involved in informational replication. Sperber has proposed a threefold test for "true" replication, and says that on the whole the transmission of cultural information does not pass. It fails his test because, when cultural information is reproduced, the copy does not inherit *from the original* the properties that make them relevantly similar. The similarities, says Sperber, are due more to the ways in which human psychological systems have evolved than to the intrinsic properties of the information.

Sperber's "test" for replication is valid and useful, but I have provided an alternative interpretation of what happens in culture. There the copy does, in my view, inherit from the original the properties that make them relevantly similar. Our evolved psychological systems are simply the mechanism via which it inherits those properties. Indeed, this is analogous to the situation in biology, where genetic information is replicated via evolved reproductive systems. My genes inherit from my parents' genes the properties that make them relevantly similar, but they depended on a mechanism in order to do so.

Of course any copying process demands a mechanism of some sort, and I have no intention of brushing the real issue under the carpet by a clever use of apparently analogous phrases. There are two key questions here, with regard to culture. First, *has* any information been replicated? Secondly, if information *is* copied, then – as Sperber has so rightly highlighted – the key issue is *from where* does the mechanism produce that information: from the original, or from somewhere else?

Sometimes, in cases which Blackmore would call copying-the-product, no information is replicated at all. A process is followed with no real understanding, and the result is a product but not the information (Blackmore's "instructions") on which the original product was based. Going back to the example of the scribble and the star, it is clear that the people trying to reproduce the scribble were engaged in this sort of process, whereas those trying to reproduce the star were, rather, following implicit instructions. What is interesting about this example, however, is that in *neither* case is any information actually copied – and this is revealed by Sperber's question about the source of the similarities between the original and the copy. From where do the copies of the star inherit their similarities to my original? The copies are based, as Dawkins says, on the (probably unconscious) instruction: "Draw a five-pointed star." Their similarities to my original are therefore due to information that the participants bring to the situation, rather than to my drawing per se. These people see a drawing of that sort of star and bring to it their existing knowledge, on the basis of which they produce a copy.

In the case of my genes, however, their similarities to my parents' are based on nothing other than the properties of those parental genes.

The reproductive mechanisms involved simply copy what is there already, without the need to interpret or add to the information that is stored in the relevant bits of DNA.

So what of cultural information? Clearly information as well as products *are* reproduced in culture, but from where do the later representations inherit their similarities to the originals? From where, in other words, do the underlying mechanisms (our evolved psychological systems) produce that information?

Sperber, Boyd and Richerson have emphasized the processes of interpretation and translation that are involved whenever cultural information is reproduced, but even if we accept this point of view it still leaves open the question of the source of the information being interpreted and translated. In cases like the earlier example of a trained engineer trying to copy a novel product, for example, Sperber is right to claim that the relevant information comes from the knowledge which he brings to the product and not from the product itself. Inference and decoding are necessary because (as discussed in Chapter 8) the product itself does not contain the necessary information. True replication has not taken place, therefore, because such cases do not pass Sperber's test.

There are myriad other cases, however, in which we derive information much more directly. If I read an article about the work of a charity with which I was previously unacquainted, then of course there is a mental process involved: I must be able to decode the written word; to understand the concepts involved; perhaps in some cases to look up the meanings of unfamiliar words in a dictionary; to infer the meaning of any unclear passages; to disentangle any agenda on the part of the writer from the true worth of the cause, and so on. Still, though, the information that I gain as a result of my reading *does* come from the article. I do not bring it to the situation. How could I, when before that moment I had never heard of that charity or its work?

If I am trying to learn how to play a new piece of violin music, then again there are mechanisms involved: I must know how to decode the musical notation; understand how that notation relates to the range of possible things I could do with a violin; have the relevant practical skills, and so on. Nonetheless, this collection of existing knowledge and skills is simply a mechanism via which I can acquire the *new* information that is contained in the sheet music. Comparable processes of interpretation and comprehension take place whenever I hear someone speaking, or observe someone using sign language. In all such cases the resultant copies derive from the originals the properties that make them relevantly similar, and the processes involved are *mechanisms* via which those similarities replicate, rather than the *sources* of those similarities. Blackmore may well have overemphasised the role of imitation in cultural replication, but this does not imply that there is no such thing as cultural replication.

Memes and the Mind

One of the issues that this debate throws up is the vexed question of the relation between memes and the mind. On the one hand Blackmore would claim that imitation is the key to cultural replication, and on the other writers like Sperber argue that it is the human mind that produces most of the similarities between cultural representations, rather than any intrinsic properties of the representations themselves. If I have trodden a convincing line between Blackmore's somewhat simplistic reliance on imitation and Sperber's rather drastic rejection of any form of replication, then what does that tell us about how active the human mind is in the process of cultural replication? Are memes effectively *self*-replicators, or are they more like "bits of replicable information"[32] which depend on the mind for their processing?

The view that has emerged from the preceding chapters is that the human mind develops as a result of acquiring memes, and many of its activities are then dictated by its memes – but that the memes themselves cannot function independently of minds, and are always initially created by a mind. Modern humans, on this story, are born with a degree of mindedness, and this is exploited by existing memes to the extent that the fully fledged mind may, itself, create new memes.

The problem with this version of events is that, so far at least, more questions are raised than answered by it. If the early basis of our mind is part of our (genetic) phenotype – a product of our brain's neonatal structure – then it must have certain features which account for the emergence of memes. What are they, and how did they evolve? If we accept that other species also engage in some form of culture, then what is so special about ours, and why are we not able to share it with them?

10

Early Cultural Evolution

Evolution cannot create something out of nothing, and it is time to ask where memes came from in the first place. What must the minds of our ancestors have been like, in order to account for the emergence of memetic culture, and what might the early evolution of memes have been like?

The Emergence of Genes

It will be useful to look first at the significant elements of the emergence of physical life on our planet. Before the emergence of replicators, there was physical material in abundance but no consistent or complex organisation. In addition to the plethora of simple matter, there were also various energy sources (e.g., the ultraviolet rays from the sun, or lightning). This energy stimulated the combination of simple matter into more complex forms. The forms that persisted would, of course, be those that were stable.

There are several theories about the type of matter that was involved in this primeval combination. Here I stick with the standard "soup" hypothesis which claims that the initial material consisted of organic molecules, but in fact it is irrelevant to the gene thesis which of the options is actually true. The important element of *any* such theory is that the most significant occurrence was the appearance of a stable form that was also able to make copies of itself. By "stable" it is meant that a particular combination of molecules will persist, if it occurs at all. If it is unstable, then it will not last for long. The result will be that some molecules (those whose combination is stable) will appear to "attract" each other. It should be noted at this point what a long shot is *any* such union. The molecules involved may collide any number of times without bonding: it is only if they collide in exactly the right place at the right time that bonding will occur.

Dawkins[1] invites us to consider what would happen if one of the stable combinations that occurred consisted of molecules which were attracted to their own kind. If just one molecule with this property should bond with, and then (perhaps because of some external influence) split from, another of its own kind, then nothing remarkable has happened. If, though, several different molecules, each of which is able to form a stable union with others of its own kind, were to come together in a particular order, then when those component molecules attracted others of their own kinds, those others would automatically arrange themselves in the same order as the original. Then, continues Dawkins, if the larger form were to split apart (if that combination were self-catalysing, say) then the original combination would, in effect, have created a copy of itself.

It does not matter whether the details of such speculation are wrong. What *does* matter is that, however it occurred, the formation of replicators would mean that the previously randomly populated matter would rapidly be filled with copies of the replicators (simply because they are making copies of themselves, whereas other forms are one-offs), and so the replicators would soon have to compete for space. Some of them would be less accurate than others, and mistakes would of course be cumulative: the ones that made poor copies of themselves would soon cease to exist. Some would be less fecund than others and these would swiftly become a minority. Some combinations would be less stable, hence shorter lived than others, and these would also (unless they were considerably more fecund) become less numerous. The result of such variation, together with the limited space, would be competition between the replicators and the dominance of the most fecund, long-lived and accurate amongst them.

Copying errors that resulted in a higher degree of stability, or in ways of decreasing the stability of a different type of replicator, would have been preserved. At least some of the combinations would become more and more complex, and those whose form provided them with some sort of protection would be at a further advantage. Some chemical combinations may have had the effect of destabilizing "rival" combinations; some may have been able to incorporate less complex combinations into themselves; some may have been able to "protect themselves, either chemically, or by building a physical wall of protein around themselves".[2]

Overall, then, the level of complexity increased through time, and the "process of improvement was cumulative".[3] In particular the methods of protection would have become more elaborate, and this eventually led to the variety and complexity of today's organic matter.

(Although Dawkins sometimes speaks as if it were inevitable, notice that it seems merely to be a matter of historical fact, supported by geological evidence, that complexity has increased over time, rather than the result of any law of nature. Indeed, it may be that our perception of the tendency towards increased complexity is based more on a subconscious belief in a Great Chain of Being than on any facts about the natural world – for it is surely the case that, once replicators were up and running, the simpler organisms continued to do rather well. Indeed, in terms of both quantity and evolutionary stability, protozoa seem to be vastly more successful than humans.)

The descendants of some such primitive replicators are our genes. Their protection devices, or "survival machines", are our cells and the bodies they inhabit. Even before this level of complexity was reached, though, the replicators had evolved to the point where the physical properties and organizational structure of their constituent chemicals produced a variety of external (structural and behavioural) effects: a "protective coat",[4] and so on. Each new generation of (accurately copied) replicators had the same effects that their "parents" had produced, since what they had in common with their "parent" replicators *was* the fact that both controlled the same effects. To put this another way: once the replicators had achieved a certain level of chemical complexity, they began to embody information *about* certain types of structure and behaviour – information with executive control *over* the relevant external features – and it was this information that was copied when they replicated.

Culture's "Primeval Soup"

I return now to the story of cultural evolution, which I take up at the point where our brains had already developed at least the sorts of capacities that we now observe in the higher apes. There is of course an unresolved debate amongst evolutionary theorists about the sequence and rate of development of the human brain, but it would be inappropriate for me to join in with this: I am not an evolutionary biologist; my interest lies in the question of *cultural* evolution, whose pace of change is much too rapid to be picked up at the level of genes. For my purposes, it does not matter whether the explosion in brain size occurred before, during or after (i.e., as cause, concomitant or effect of) the initial emergence of cultural replicators.

The starting point, in keeping with the biological story, should have been a plethora of simple mental material. Yet it is not so easy, in the case of the mental, to see in what this might consist. "Material" is not, in fact, a helpful word in this context, since it has overtones of the mysterious "immaterial substance" of the Cartesian soul. It may be better to talk of our ancestors' mental *activity*, although it should be borne in mind that this term is intended to include their potential activity given their abilities, as well as their actual activity at any one time. If it is, as most would concur, acceptable to refer to the kinds of things that go on in the brains of the higher mammals as mental activity, then the brains of our primitive ancestors would also have provided a world of simple mental activity.

So the elements in the primeval mental "soup" consisted in primitive mental activities. The next step in the physical story was the emergence, under stimulus, of stable forms and thus of a higher degree of complexity, without which replicators could not have evolved. In order for cultural replicators to develop, an increasing degree of complexity and stability would have been needed in that realm, too.

Although we ought not search in the cultural realm for analogues of the finer details of the physical story, if a plausible parallel may be drawn between the two then it should not automatically be ignored. It may be that the development of complex stability in the cultural realm did develop along lines parallel to those along which it ran in the physical realm. If so, then the story told below may be true as well as illustrative. Recall, though, that the physical story itself is necessarily speculative, so it may be that neither is strictly accurate. In both cases, the point is that evolution could not get going without some mechanism for the emergence of complex stability, and so we have to be able to tell some plausible story about that mechanism. If, one day, we are able to discover what the *true* story is, then so much the better, but for the moment what matters is that some story can be told – and since one of the plausible stories in the cultural realm parallels one of the plausible stories in the physical realm, those two stories have an added attraction and have been used here.

The stable union of two behaviours will obviously involve something very different from that of two molecules. In the case of the behaviours, which are functional rather than spatial in nature, it must mean that they form a union in time rather than space, and I take this to be a cause-effect relationship. Just as, when the orientation is right, the properties of one molecule cause another token of its molecular type to *be next to* it (spatial relationship), so in the right circumstances we should expect the properties of one behavioural token to cause another to *happen just after* it (temporal relationship).

As an example, a common observation amongst modern animals is that rewarded actions are repeated whilst those that are punished are not: such behaviour is said to be governed by the "law of effect". Now, if B sees A engaged in some behaviour that is rewarded, and B himself is already capable of the behaviour, then he is likely to mimic A so that he too will be rewarded. This is the sort of social learning ("stimulus" or "local" enhancement) that Blackmore and others would describe as the "reproduction" of behaviour. One instance is the behaviour of a blue tit which pecks at a milk bottle top because it has seen other blue tits doing the same thing, as discussed in Chapter 9. Such learning is certainly observed amongst today's animals, and the assumption of its availability to early hominids is thus unproblematic (remember that I am not claiming to offer a full account of their evolution, but am taking up the story at quite a late stage).

Such a process may be redescribed as follows. Two creatures are both capable of a certain behaviour – behaviour which, as it happens, falls under the law of effect. When one of the creatures manifests its ability in the relevant behaviour, the other creature is moved to do the same. Or, to put it another way: two mental activities have come into contact with each other in just the right circumstances, and the manifestation of one has followed – been temporally "connected" to – the manifestation of the other. This gives rise to the appearance of "attraction" between the two tokens of that behaviour, just as it does when two tokens of a molecule become spatially connected.

It is also in keeping with the molecular analogy that neither of these behaviours is, in these circumstances, a genuine replicator: this reflects Blackmore's description of such learning as "reproduction" rather than genuine "replication". (Similarly, when molecule m_1 joins with and later breaks away from molecule m_2, it is obvious that m_1 has not *created* another token of an m molecule.) If the shorthand "behaviour" is used, then it does seem that A's behaviour has created another token of its type: when B sees A, he repeats A's behaviour. If, on the other hand, we use the more accurate "mental ability that gives rise to behaviour", then it can be seen that A's token has merely, when manifest in the presence of B's, led to B's also being manifest in behaviour. A and B both already possess the relevant mental ability, and therefore A's token of its behavioural manifestation does not need to *create* another. When (via A's behaviour) it comes into contact with B, it simply "attracts" B's own token of its type (causes it to be manifest soon after A's).

It might seem obvious at this point that, since many modern mammals are capable of (at least limited forms of) this sort of behaviour, there is nothing here that could account for the emergence of the uniquely human phenomena of memes. On the other hand, the fact that the primordial chemicals may still be found today does not imply that some of the original combinations of them did not develop into genes.

Behavioural Patterns

The difference between the chemical combinations that evolved and those that did not lies in the *effects* produced by the particular formations: the effect of the ones that were composed of "self-attracting" molecules was that they made copies of themselves. One of the important features of the physical story is that, in the replicable patterns of molecules, the union between the building-block molecules is more stable than the union between each of them and other tokens of its own type – this is why they maintain their union with each other, whilst breaking apart from the "copy".

In the mental world, the "building block" molecules are replaced by mental activities that give rise to the reproduction of (or apparent attraction between) certain behaviours. Now, are there circumstances in which – just as in the physical story such molecules can form stable unions with each other – tokens of these types of mental activity might come together in just the right order at just the right time, so as to form stably bonded behavioural patterns? And if such patterns *were* formed, of behaviours which naturally give rise to their own repetition, then would those patterns analogously begin to make copies of themselves? Clearly, this can only happen if the organisms involved are capable of both imposing some sort of organizational structure on their own actions and learning such patterns from each other.

The difference between low-level copying and this more complex sort of learning has been untangled by the psychologists Richard Byrne and Anne Russon.[5] In a discussion of different levels of imitation, they distinguish between "copying the organizational structure of behaviour" and "copying the surface form of behaviour". Copying the "surface form" of behaviour is the lower-level activity, in which the fine details of particular actions are imitated. Infants do this all the time when they attempt to reproduce things they have seen their parents do, like mop up spilt drinks or build a brick tower. In order to go further and copy the "organizational structure" of another's behaviour, an organism must be able to arrange its existing behaviours into new (and potentially complex) patterns. The component actions within those patterns may well vary from individual to individual, being developed via trial and error; but the overall structure of the behavioural pattern must be fixed if it is to count as this sort of imitation. At this level, an organism is copying "the structural organization of a complex process (including the sequence of stages, subroutine structure, and bimanual co-ordination), by observation of the behaviour of another individual, while furnishing the exact details of actions by individual learning".[6]

Byrne and Russon characterize their approach to imitation as "hierarchical", and indeed it is reminiscent of Chapter 4's discussions of the assembling constraints on complex replication. In particular, it echoes Koestler's claim that a set of invariant rules will account for the structure and stability of complex assemblies of information or behaviour, with variation allowed in the "strategies" that are actually employed (to illustrate, I used the example of chess, where the rules are fixed but what happens during any one game will vary considerably). The replication of complexity always depends on such a hierarchical structure. Thus my suggestion that memetic replication was preceded by the emergence of stable behavioural patterns (analogously to the early days of pregenetic replication) is strongly supported by Byrne and Russon's thesis. They say, in effect, that the ability to copy such a pattern relies on the ability to pick out which of its elements are fixed "rules" and which are variable "strategies" – in other words, to impose a hierarchical structure on the complexity being copied. Thus the early replication of simple behavioural patterns would, since it is a hierarchical process, involve exactly the feature that would best support the future replication of complex information (memes).

There is also a link between this understanding of how behavioural patterns are learnt, and Terence Deacon's claim[7] (discussed in more detail in the next chapter) that a vital feature of modern human infants' minds is the ability to see structure beyond the details of natural languages. If the capacity to learn organized patterns of behaviour depends on being able to pick out a pattern's general structure from the details of its particular instantiations, then again this is highly significant for memetic evolution. It means that the early replication of simple behavioural patterns necessarily involved a feature that would later facilitate the replication of much more complex information.

In summary, it appears that the capacity to organize one's behaviour into patterns, and to learn those patterns from others, is a hierarchical process that depends on the ability to pick out the structure of a pattern's organization. Even when it operates at a fairly primitive level, therefore, such a process has the *potential* to support the replication of much more complex and abstract representations, should such representations be available to the creatures involved.

An Example

My claim, then, is that in the primordial mental "soup" of our ancestors' mental activities, a proportion of those activities would have had the crucial property of "attracting" other tokens of their own type – and that, given the right stimuli, *patterns* of such activities might have begun to form.

In support of this claim, it will be helpful to explore an example of some "self-attracting" behaviours which, if they were to come together in just the right place at just the right time, would form a stable pattern of activity. Once more, of course, I am not suggesting that my example is a true picture of the emergence of the first cultural replicators: the less so in this instance, since it concerns particular instances rather than general patterns of behaviour. Particular instances can, though, be used to illuminate general truths, and in this case I hope that some light may be shed, by a fictitious example of the way in which mental replication *could* have begun, on the general truths about that process as it was *actually* initiated. The story offered will have no pretensions to being an account of the origin of memes, but will nonetheless be useful in providing a consistent illustration to which I can refer throughout this chapter's discussions.

An appropriate example would revolve around two distinct types of behaviour, each of which involves mental activity and is "self-attracting". For the first, I have chosen the use (well documented in modern primates) of a stick to "dip" ants out of their nests so they may be eaten. This behaviour is learnt rather than genetically hard-wired, and displays clear signs of mental activity (finding the ants' nest, finding a stick, controlling the digits to manipulate the stick, etc.). The second is the much simpler act of stripping the leaves from a bush, in order to eat them. This, I assume, is also a learnt rather than a genetically blueprinted ability. Both are also likely to be "self-attracting": if A finds an ants' nest and is seen to be obtaining food from it, then, in the right circumstances, B would soon join in. Similarly, if A is seen to be feasting from a particular bush, then B is likely to join him.

The "right circumstances" are important, however. Just as, in the physical case, for a molecule to bond with another it must be oriented in exactly the right way at exactly the right time, so in this case A's behaviour must take place in the "right" circumstances if its "self-attracting" properties are to be realized. In particular, B must be in the vicinity, aware of A's activity, hungry, and already capable of the activity involved. In such circumstances there is nothing mysterious about the "self-attracting" property of such behaviours: it is largely a product of the fact that they are rewarding. Nor is there anything mysterious about the stable pattern that might be formed from the combination of such behaviours, which would obviously be the act of stripping leaves from a twig in order to eat them, and then the use of the twig for ant dipping.

Innate Prerequisites for Primitive Replicators

Yet if an organism is capable of copying not only the simple activities of its conspecifics, but also the organization of their more complex behaviours, then why shouldn't we say that it is acquiring cultural replicators (albeit fairly primitive ones) – and if so, then why don't we just call *them* memes?

I think that it does mean that such behaviours are the results of *primitive* cultural replicators – but I don't call them memes because there is a crucial distinction between the two sorts of replicator, which I expound in later sections of this chapter. Nonetheless, it is worth asking what were the mental capacities which enabled our ancestors to participate in even such a basic process of cultural replication – for it was this existing process which provided the foundations for the development of memes. What were the innate prerequisites for the emergence of the simple replicators on which memetic evolution was inevitably dependent? I shall not pretend to offer a comprehensive description of homo's mental life, but do need to specify the elements that would have been necessary in order for replication to arise.

It is reasonable to assume, throughout the following discussion, that the capabilities of modern-day primates are a good pointer towards the abilities of our ancestors at the time that their minds were emerging. This is not to make the mistake of thinking we are descended from our closest living relatives, the chimpanzees: we and chimps have a common ancestor, but it is very likely that chimps are as different as we are from the creature that preceded the fork in our lineages.* Nonetheless, that fork is relatively recent, and it seems likely that if our mind-developing ancestors were relatively big brained and socially organized, then their abilities may be reflected in those of chimps. "We must beware of equating chimpanzee cognitive capacities with those of all African apes, including the ones of five million years ago. Cautiously, however, I suggest that we can say that large-brained apes that live socially complex lives are likely to develop a chimpanzee level of consciousness."[8] Putting this another way: if there are basic faculties that human babies share with chimps, then it seems likely that the "intermediate" ancestors would also have had them.

The most obvious prerequisite for copying is an awareness of one's own activities. Few people would be willing to characterize modern primates as having "consciousness" in the human sense, but there is evidence – both anecdotal and experimental – that at least some of the higher primates are aware of their own activities. Gallup's controversial "mirror test"[9] is one example: chimps, once familiarized with a mirror, have a red spot marked on their head – shown the mirror anew, they (and orang-utans, but not gorillas) touch the red spot on their own head, demonstrating that they recognize the image as their own.

Perhaps the most convincing demonstration of the fact that some primates are aware of both their own and their companions' activities, however, can be found in the documentation of deception. Although mostly anecdotal, there is widespread agreement amongst primatologists that both apes and some species of Old World monkeys (e.g., baboons) have frequently been observed to partake in "such tactics as concealment, distraction, the creation of misleading indications of intent, and manipulation of innocent bystanders".[10] In one simple instance, a male chimp who was about to engage in a confrontation with another "was baring his teeth in a fear grin", and then he pulled "his lips over his teeth, wiping out the fear-grin. He did it several times. In mutual intimidation between males, it makes sense to hide signs of nervousness. That's what [he] seemed to be doing."[11] There are many more instances that could be quoted; what

* Moreover, as Steven Pinker has pointed out in *The Language Instinct* (1994: 342–9), our assumed "closeness" to chimps and gorillas is largely the result of extinction: if some of the intermediate species (i.e., the hominids that came before modern man but after the time when our line forked from chimps) had survived, then the chimps would not seem to be so remarkably close to us. Conversely, if all other apes were extinct then maybe monkeys would be the creatures that were chosen so arbitrarily to be the focus of our research.

matters about them all is that, in order to practise deception, an agent must have a fairly clear picture of both his own and the deceived creature's intentions and activities. Both capacities are necessary conditions for the emergence of mental replicators, since it would be impossible for A to copy C unless he were aware of both C's and his own actions.

Nonetheless, it should be noted that the evidence from primate deception is open to interpretation, and it may plausibly be argued that it demonstrates an awareness only of others' actions and perceptions, not of their intentions. Thus "Whiten (1993) points out that there is as yet no experimental evidence for false-belief attribution in primates." If it is adequate, for the purposes of tactical deception, to be "adept at controlling what is perceptually available to conspecifics",[12] then primates will be able to practise such deception even if they have little awareness of others' beliefs or intentions. Without an awareness of the actions and perceptions of herself and the deceived, however, it is certain that a deceiver could not practise her art – and equally certain that mental replicators could not emerge. The extent to which A would need some sort of theory of mind in order to mimic C may be open to question, but his need to be aware of what both are doing and perceiving is not.

Clearly, though, these capacities alone are not sufficient for replication. In order to be able to copy C, A must not only be aware of C's behaviour and his, but also be able to *link* the two – and in order for this to happen, there must be what is known as cross-modal sensory integration: the capacity to make sense of information from different sensory modalities (sight, sound, touch, etc.). Is there evidence to suggest that this ability is innate in either humans or primates? Classically, the debate has been flanked by Piaget, who claimed that cross-modal integration is not innate but achieved during development, and Bower, who said that early perception is supramodal (i.e., "the sense modality of the inflow is disregarded")[13] and development renders the senses distinct. Current evidence suggests that neither hypothesis is quite right, but that both human and primate neonates are capable of primitive cross-modal integration, and that development enhances it in humans but not in primates.

The classic experiment[14] involves two types of dummy (pacifier), one smooth and one nubbed. Human neonates were found to be capable of matching a visually perceived shape to that which they had previously explored tactually, which indicated that (at least a certain type of) cross-modal integration was available without having to be learnt. Similar experiments on infant macaques[15] gave corresponding results. Cross-modal abilities were observed during the course of the latter experiments in infants as young as eight days old (roughly equivalent to month-old humans). Such experiments suggest that both human and some primate neonates are "capable of using and storing surprisingly abstract information about objects in their world. This information must be abstract enough, at least, to allow recognition of objects across changes in size and modality of perception".[16]

If such abilities are now innate in humans and – perhaps more relevant to the current project – in primates, then it seems plausible that our ancestors would have had them too. So A is aware of his own and of C's activities, and is capable of integrating them across their differing sensory modalities. Yet why should he do this? Unless he has some limited form of means-ends reasoning – the ability to manipulate representations of non-actual states of affairs – then there is no reason why he should bother to link C's activities with his own potential ones. Is there any reason to suppose that he may have such an ability? Once more, the capacity for deception can be invoked, since it provides a clear demonstration of this ability. Unless he were able to consider the consequences of his actions then it is not obvious how a creature could put to use his knowledge of his own and his companion's activities.

There is still a missing element. A can recognize and match his own activities to C's, and his imagination will allow him to "try out" the relevant behaviour for himself. Yet, if it is successful, then he has also to remember it: he needs, in addition to the other abilities specified, a long-term memory for facts. This, too, can be seen in modern primates, where it is demonstrated most clearly in their capacity for complex social interactions. Much of a primate's life is spent in nurturing his own and assessing his rivals' alliances: "If alliance networks were permanent structures within a troop, it would be difficult enough for individuals to cope with their intricate connections. But they are by no means permanent. Always looking to their own best interests, and to the interests of their closest relatives, individuals may sometimes find it advantageous to break existing alliances and form new ones, perhaps even with previous rivals. Troop members therefore find themselves in the midst of changing patterns of alliances, demanding yet keener social intelligence to be able to play the changing game of social chess."[17] In order to keep track of the shifting pattern of "friendships", each primate has to have a long-term memory in which to store facts about who is allied with whom.

In conclusion, then, the prerequisites for the emergence of mental replicators are likely to have been (at least limited forms of): an awareness of the creature's own and her companions' activities and perceptions; the ability to link the two; a degree of means-ends reasoning to tell her why she should *want* to link them; and the capacity to remember the sequence of events that consequently occurred. Hierarchical considerations indicate that there would be severe limitations on the content of what could be replicated, unless the creatures involved were also capable of copying the organizational structure of behaviour: they would need *sufficient* means-ends reasoning to enable them to pick out the significant features of a sequence of activities (to see the *point* of what they observe); and sufficient memory power to keep those features in mind when reproducing the sequence themselves.

It is plausible that homo should have possessed these capacities, since all are demonstrable in modern primates as well as humans. This is not, of course, to say that this list should be regarded as a comprehensive summary of the emergent minds of our ancestors; it is merely to claim that it comprises the most significant aspects of the *necessary* conditions that their mental abilities must have met in order for primitive mental replicators to emerge.

What Is Special About Memes?

Yet if I have been successful in sketching the sorts of capacities that creatures would need to engage in cultural replication, then surely I have in the process undermined the memetic project. It is usually taken to be a corollary of the meme hypothesis that memes are a uniquely human phenomenon – but if both early hominids and some modern primates are able to construct replicable patterns of "self-attracting" behaviours, then what is so special about human culture? To rescue memetics, I need to be able to explain both how memes may be distinguished from other types of cultural replicator, and how humans can be distinguished from other types of replicating creature.

In Chapter 3 I argued that, just as genes are based in DNA, so memes have their basis in representational content. In the course of that discussion, it emerged that there are different sorts of representation: some are so simple that their content may even be indeterminate; others are more complex, and their content is partially determined by the internal links that are formed between them and other representations. Now I should like to add a third level of representation: those whose internal links enable them to represent, not external objects or events, but *other representations.* These, philosophers would call meta-representations. It is at the level of meta-representation, I would argue, that memetic replication emerges.

A creature that is capable of copying a pattern of behaviour must have some sort of representation of how to perform those behaviours. There must even be internal links between some of its representations (e.g., between its representation of how to strip leaves from a bush and its representation of how to dip for ants), which enable it to produce a constant pattern of behaviour. What it need not have is any understanding of the point of what it is doing. It is quite possible, for example, that a primate engaged in the leaf-stripping/ant-dipping pattern of behaviour has no more understanding of why the ant dipping should follow the leaf stripping than a parrot has of the meaning of the words that it utters. If so, then this primate's behaviour is "complex" only in the sense that it entails a whole series of activities: although there may be *more* mental processes involved in its representation of the behavioural pattern than in its representations of the pattern's component activities, essentially the same *sorts* of mental process are involved in each case.

In order to understand the point of its behavioural pattern, the creature would need to be able to think about what it is doing – and in order to do this, it would need to be able to form a representation of that behavioural pattern. I don't mean a representation of how to perform the behavioural pattern (it already has this), but a representation of the *pattern* itself. There is a certain structure to its representation of how to perform that pattern, and it is this which the creature needs to recognize, if it is to be able to reflect on its own actions. Only then could it begin to understand that the crucial link between the separate elements of the pattern is the creation of an ant dipper. Only then, in fact, could it form the concept of an ant dipper – a concept that will emerge from its meta-representation of the behavioural pattern.

The distinctive feature of memetic replicators, then, is this abstract conceptual quality, which comes from the human ability to meta-represent. Only once we could meta-represent was it possible to manipulate and reflect on our representations independently of their original context. I would strongly argue that primates do not have this ability. "In this connection the classical experiments of Madame Kohts of Moscow were illuminating. Some of the apes observed by her used sticks as levers, for digging up hidden objects or for extending the reach of their arms. Her chimpanzee would pull a loose board from a case and use it, but if the case's surface was unbroken he could not see a possible stick in it."[18] These apes must have had a representation of how to use a stick as a lever, but they were unable to reflect on that representation: they could not meta-represent the more *abstract* concept of a lever as an independent object, transferable between contexts and activities.

Indeed, to what extent their "concept" of a stick is accurately to be described *as* a concept is open to debate. Although they had a perceptual grasp on a stick as an object in their environment, and the imaginative ability to make use of it in certain contexts, this is not usually all that is meant when we say that a creature has a concept of something: a concept is more than a discriminatory, context-dependent ability. It is a more abstract representation, universalized rather than task specific, available for use in many different contexts. It is, in other words, a meta-representation.

In order to distinguish clearly between the two types of concept, I shall from this point refer to representational concepts as *notions*, and reserve *concept* for meta-representational concepts. My claim, then, is that it is these more abstract concepts that were crucial for the emergence of memes. Only once a creature can *meta-represent* (give labels to its representations, and manipulate those labels in its mind) can it lift its representations out of their original context and use them in another arena.

Yet non-human animals do not seem to have made this leap, despite the fact that at least some are able to copy each other – to engage in primitive mental replication. What enabled our ancestors to begin to meta-represent, in the process unleashing the power of memes?

Innate Prerequisites for Concepts

Brain Size

The most obvious difference between human and nonhuman primates lies in the size of our brains: it is estimated that even the earliest homo had a higher encephalisation quotient† (EQ) than modern apes. It is arguable that this was the result of a cycle of positive feedback, triggered when hominids expanded their diet to include meat as a regular rather than occasional foodstuff. As Dawkins puts it: "In the mammals as a whole, carnivores typically have a slightly higher EQ than the herbivores upon which they prey."[19] Throughout evolutionary history, the successful species have been those with the optimum trade-off between their beneficial features,

† An animal's encephalisation quotient is the ratio of its actual brain size to the expected brain size for an animal of that size.

given the resources available – and the general message about brain size seems to be that any species will have as big a brain as it can afford. This means that if a species needs its energy to escape predators or to chase prey, then it will be a waste of that energy to spend too much of it on the brain. Conversely, the brain is so powerful and useful that if a species has a stable environment and plentiful supply of nourishment, then its members' brains will continue to grow.

Now, early homo was bipedal, with free forelimbs that terminated in hands. He was already able, because of this, "to thrive where an ape could not live",[20] and later he took advantage of the evolutionary opportunity afforded by this potential for technology and started on an embryonic form of the hunter-gatherer lifestyle. "But evolution is rarely simple cause and effect. There are many variables in the uncertain mix: the climate, the local geography, a species' evolutionary heritage, the nature of other species in the community, and a measure of pure chance."[21] Probably the best we can say about early homo is that a combination of such variables, including the stable environment and plentiful food supply afforded by his relative fitness to the new climate, meant that there was no need for a trade-off between brain expansion and other adaptations.

The question for memes is whether the "great encephalization" would have been enough to enable our ancestors to begin to meta-represent. Perhaps the "spare" brain power was all that was needed for this new cognitive capacity. Alternatively, it may be that a radical divergence was needed from the neural architecture of their ancestors before hominids were capable of forming representations of their own internal representations. My guess would be that the emergence of the ability was closely allied (again, I don't really want to become embroiled in the debate about whether this was as cause, concomitant or effect) with the explosion of brain power, rather than with some novel cognitive architecture. This is because, in order for memes to take off as they obviously have done, the vital facility has to have been practically universal – otherwise, the emergence of fully fledged concepts in one of our ancestors would have had little to no effect on his contemporaries. A concept cannot replicate if none of the surrounding organisms is receptive to it: household pets do not develop a humanlike capacity for culture simply as a result of exposure to it, for they are not appropriately receptive.

My search for the origins of memes is not, however, dependent on a parallel search for the neural architecture that facilitated their emergence. I want to know what *mental* abilities enabled our ancestors to begin to meta-represent when their primate cousins could not – and this is a quite separate issue from the question what neural structures and functions underlay such abilities.

Comparing Representations
The most crucial facility for this purpose is the ability to make links between one's representations. Yet I have already argued that any creature capable of associative learning must be able to do this, so what is there here to facilitate the emergence of meta-representation? The vital next step is for those links to be freed from the representations which initially triggered their formation – and this depends on the ability to *compare* one's representations. Only then will it be possible to form representations of the links between them.

Concepts (of the fully fledged, meta-representational sort) play a much wider causal role than that played by notions. Since notions' effects are restricted to a particular context, their causal roles (i.e., the range of things that can cause them and which they in turn can affect) are bound to be limited. Thus our hominid's representation of the stick that he uses for ant dipping may properly be called a *notion* of a stick: it is triggered by the sight of a stick (either on the ground, or in a bush as revealed by leaf stripping), and in the right circumstances it will have control over his ant-dipping behaviour; but that's it. Concepts, in contrast, have a more extended causal role, whereby they may be triggered by a variety of stimuli, and can control a variety of actions. Had the hominid the *concept* of a stick *as an ant dipper*, then its causal role would be freed from the original context, and he might for example actively set out to *create* an ant dipper by stripping leaves (even if he doesn't intend to eat the leaves), or take his ant dipper with him in search of termites.

Now, the hominid could not have learned this behavioural pattern in the first place without the ability to link his representations of how to strip leaves and dip for ants. What he needs now, in order to form a concept of an ant dipper, is to compare his representations and "see" what they have in common: to form a representation, in other words, of the link that he has already made between the two. He needs to recognize the fact that the two activities are linked by the stick (or more accurately its dipper properties – being long, thin, etc.), and in so doing he will free his representation of that object from its original context as a bridge between two other representations.

Indeed, the establishment of a concept will always initially be a comparative process: we need to be able to compare our notions, in order to abstract their important common features from the various contexts.

This is reminiscent of the fact that an important part of any replicator's content is determined by the *differences* between itself and its alleles. What we humans can do is to *see* those differences, and to form representations of them – and as soon as a creature can compare its representations with each other, its mental life loses its dependence on the external world. A creature whose representations are tied to a limited range of external stimuli and actions cannot manipulate those representations in a different context from that limited range. A creature whose representations have been freed from those bonds, however, may manipulate its concepts quite independently of their original contexts. No longer do its representations depend on external stimulation: representations of links between representations can be triggered by the original representations themselves, whether or not the creature finds itself in their usual context.

Meta-Representation Today

My claim that the emergence of concepts is always a comparative process gains credence from a consideration of how meta-representational concepts emerge, even today. Few people have any clear recollection of the way in which they first grasped concepts such as "sheep", "tree" and other basic elements of our understanding of our surroundings. We can, though, be clearer about our more recent acquisition of more complex concepts such as "cantilever", or "harmonic minor scale". They appear to be acquired in two steps: at first the notion will be a peripheral part of a theory or activity, the bulk of which is familiar. The reason why it is peripheral is that our understanding and use of it is wholly context dependent. We are not able to speak of that entity outside the circumstances in which we usually encounter it, for our interaction with it depends on our interaction with its context. As an example, consider the concept of a cantilever. Many people have seen the Forth Bridge; plenty of women wear a bra; in most of our houses there are shelves. A cantilever is therefore a structure familiar to most, yet without some additional stimulus many people will never acquire its concept (meta-representation).

The second step – the step that will disentangle the relevant word or entity from its usual circumstances – will be the combination (in varying proportions) of its familiarity, and an appropriate stimulus. The more often we encounter a novel subject, the more of its features we shall appreciate; if there is, in addition, an appropriate stimulus (e.g., a pressing need, or some form of hint), then eventually we shall be able to compare our individual representations and extricate their common features. In the case of the cantilever, the stimulus might simply be a definition, encountered in a book or conversation. It might, though, take the form of a practical need: if a person is building something, and trying to find a way of supporting a structure within it, then this might prompt her to consider the form of similar structures with which she is already familiar. If she abstracted from those structures the common, significant feature that it is fixed at one end only, then she could apply it on her own construction, away from its usual context.

As another illustration, think of a novice violinist who is able quite competently to play several major scales, as well as the arpeggios of C and G major. If she is asked to play the arpeggio of D major, then she will not initially know what to do. In order to produce it, she will have to think about the arpeggios she knows, in an attempt to ascertain their common features. Once she has abstracted the rule governing the sequence of intervals in an arpeggio, she will be able to apply that rule to the new situation: now she should be able to produce the arpeggio of D major, and indeed of any other key for which she is asked.

Clearly, the sorts of concepts that first emerged would not have had anything like the complexity of arpeggios or cantilevers, but it is not implausible that they should have been acquired by a process which was in essence similar to that described. The subject of the concept moved, in other words, from context dependency to abstraction as a result of our being stimulated to compare the common features of various familiar notions.

Thus the key element in the emergence of concepts was the facility for internal comparison of representations, without which it is impossible to discern their common (functionally relevant) features. A vital consequence of this would be the ability to abstract information from incoming representations as well as from the environment – and this faculty would have been useful enough for selection to favour those of our ancestors who were not only able, but also *tended* to represent and to compare representations. (Indeed, this view is supported by observations of variation in modern humans' tendency to make internal mental links, and the fact that a bias towards this tendency is a key characteristic of gifted individuals.)[22] It would thus have provided an efficient basis for the emergence of the new form of evolution – and combined with the sorts of mental abilities that have been postulated as the precursors of replicating notions, this would have enabled our ancestors to develop communicable, manipulable, memorable and widely applicable representations: i.e., replicating concepts, or memes.

Which Replicators Count as Memes?

The implication of this chapter has been that a clear distinction can be drawn between primitive cultural replicators and the memes that pervade our culture today. Other primates, including our ancestors, do clearly engage in some sort of cultural replication, but if humans alone engage in meta-representation then we ought to reserve the term "meme" for this specialized type of cultural replicator. Yet obviously this view of memes, as replicating meta-representations, is not the only way of characterizing them. Susan Blackmore would argue that memes' distinguishing feature is not that they are a particular sort of cultural replicator, but that they are *replicators* at all. She is clear that some copying methods do and others do not involve true replication – and that only imitation can support memetic evolution. This section defends my hypothesis against her challenge, and clarifies the place of memes in human culture today.

Imitation Again

Blackmore points up the difference between the reproduction in a novel context of behaviours of which the copier is *already* capable, and the genuine imitation of novel behaviours. She uses the case of a blue tit, learning to peck milk bottle tops, as an example of the reproduction of behaviour: in her view the blue tit has not acquired a meme because it has not really imitated anything. This is in contrast to my view that the question whether the blue tit has acquired a meme will be resolved not by a study of the *method* of information transmission (was it really imitated or just reproduced?) but by an examination of the information itself (was it a meta-representation?).

Blackmore would concur that learning *is* involved here: the blue tit has learnt where to seek food, even though it has not learnt *how* to do anything. Now, it is not immediately obvious to me that information about location cannot constitute a meme: compare "you can get food if you peck these objects," with "you can get food if you shop here." Surely the relevant question is not whether a new skill has been acquired – as opposed to merely new information about where or when to apply an existing skill – but whether the information acquired is of the right sort to constitute a meme. I would argue that the blue tit cannot engage in memetic replication unless it is able to develop context-independent meta-representations.

In the blue tit's case, its understanding of milk bottles is arguably untestable. There would be no observable difference between two birds, one of which was simply engaging in associative learning and the other of which had genuinely acquired a novel concept. The problem in the case of this example is that it is hard to know what it would mean for a bird to have a concept of a milk bottle that is "transferable between contexts": in what other context could a milk bottle be of interest to a bird, other than the one in which it contains milk that the bird can reach by pecking? And without a separate context in which to "test" the birds, there is no way of discovering their true understanding of the situation.

In any case, although it is at least hypothetically possible for the bird to have formed a concept of the milk bottle, in all probability Blackmore is right to say that blue tits do not engage in memetic replication. The reason for this, however, is that they are unable to form complex, context-independent representations; it is not that no new skill has been acquired. The question whether memetic replication is involved in a particular instance of social learning will not be answered so much by analysing the *method* of information transmission (reproduction or imitation), as by looking at the *content* of what is transmitted (notion or concept).

Different Levels of Imitation

If the crucial distinguishing factor in memetic replication were simply the method of transmission (imitation or not), then we should expect to see a clear cut-off between imitation and other forms of social learning, but in fact this not the case. The blue tits provide a hypothetical example in which a creature's observable actions may be interpreted in two quite distinct ways. Blackmore is right to say that what has been learnt is not a novel behaviour but simply information about a new situation in which existing behaviour will produce rewards – but she is mistaken to think that this alone can tell us what has really gone on in the blue tit's brain. There is no way of telling by behavioural observation whether the information has been represented in a limited notional form, or whether it constitutes a more complex, context-independent concept. The blue tit has not imitated a novel behaviour, but (hypothetically) it may nevertheless have acquired a meme.

In Byrne and Russon's distinction between different *levels* of imitation ("copying the organizational structure of behaviour *versus* copying the surface form of behaviour"),[23] as in my characterization of memes, the differences are determined by the *content* of what is imitated. Whereas copying the surface form of behaviour involves a relatively mindless imitation of each individual detail, it is not possible to copy an activity's organizational structure without the ability to think a bit more deeply about what is going on: to distinguish between the fixed "rules" and variable "strategies" of that behavioural pattern.

What I have introduced, with my emphasis on meta-representation, is a third level of understanding: the capacity to recognize what has been imitated. It is this which gives rise to a representation, not only of how to perform the behavioural pattern, but of the pattern itself: a meta-representation of what has been imitated. What is significant to the debate about how to characterize memes is that the three separate sorts of process – imitating the details of individual actions, copying the structure of a more complex behavioural pattern, and meta-representing that structure – can operate at different levels within any given behavioural structure.

If Janet wishes to learn how to change a car wheel then she can absorb that information in a variety of ways. There are several stages to the procedure – jacking up the car, removing the old wheel, fitting the new one, and so on – and within each stage there is a collection of detailed actions to perform. Watching and learning from Neil at her local garage, Janet might simply copy each of his detailed actions without any real understanding of what is going on. When reproducing them herself later, minor variations will almost inevitably be introduced, and if she has failed to grasp the point of each bit of the routine then her deviations from Neil's method may be disastrous. She might, for instance, have observed that Neil tightened the bolts progressively in a star pattern, but not understood that this mattered: if she simply tightens each bolt fully, in turn, then damage could result.

Janet might, on the other hand, have copied the structure of Neil's actions, so that the changes she introduces will stand less chance of being dangerous or damaging. In this case, she has picked out the important elements of the process, and any variations that she produces will be insignificant as far as the end result is concerned. So, for example, Neil may have pushed his metal tool box under the car before fully removing the old wheel (so that if the jack slipped then the car would fall onto the box rather than the ground, and it would still be possible to reinsert the jack), and Janet might put the new wheel into that space instead – fulfilling the same purpose in a different way.

At a third level of understanding, Janet might both have copied the structure of Neil's actions, and thought a bit about that structure – and then the variations that she introduces may even be improvements. Neil may have tightened the bolts with an impact wrench, which indicates that for him the question what tool is used for this purpose is a strategic detail: what matters (the "rule") is that the bolts are tightened to a point where the wheel is securely attached; it must not be under-tightened. If Janet later reflects on the procedure that she has learnt, then she may come to wonder whether it is possible to cause any damage by *over*-tightening the bolts – which indeed it is. A better method than Neil's would be to tighten them only so far, before finishing off with a torque wrench. Now, if Janet introduces this variation then she will have made a change in the very structure of the wheel-changing process. Whereas the structure that she learnt from Neil included the rule "don't leave the bolts too loose", the structure that now governs Janet's behaviour includes the rule "make sure the bolts are neither too loose *nor too tight*."

An interesting fact has emerged from this example about the sorts of variations that will be introduced, depending on the level at which imitation has taken place. If Janet manages accurately to copy the structure of Neil's actions, but does not take the time to reflect on what she has learnt, then any variations that she introduces will involve only the relatively trivial "strategies" that intersperse the more significant "rules" of the process. If, on the other hand, she simply imitates Neil's detailed actions, then her failure to distinguish between rules and strategies might lead her to make a significant mistake, because she does not realize which bits of the process are structurally important: her variations might occur in either the detail or the structure of the process. This is also true of the situation where she both copies the structure of the behaviour and forms a meta-representation of that structure: here again her variations might occur in either the detail or the structure of the process – but in this case any structural variations will be deliberate and considered, and may even be improvements on the original.

Our leaf-stripping, ant-dipping hominids may also, of course, have a variety of levels of comprehension of their actions. It is perfectly possible that, although A has really grasped the concept of a stick as dipper, B's imitation of A is relatively mindless, copying the "motor action details"[24] rather than their functional arrangement – or even copying the structure of A's behaviour without the ability to reflect on it. In neither of these cases would B have acquired a concept at all: his behaviour would still be context dependent (having stripped leaves from a twig, he now has a tendency to break it off and search for an ant's nest) even though it has gained in complexity. The move to conceptual, structure-level comprehension is not an additional learning task but rather a matter of insight, involving "a recoding of previously available but unlinked bits of information".[25]

Memes and Modern Human Culture

If imitation is, as Blackmore would have it, the key to memetic transmission, then the significance rests on this *level* of imitation (the push to conceptual understanding) and not on whether a new skill is acquired as opposed to a new context for that skill. New skills – even ones that seem quite complex – can be acquired via the mindless reconstruction of detailed motor actions, or even the relatively mindless reconstruction of structured behavioural patterns. If human culture consisted in nothing more than this, then it would hardly merit memetic study. It is only when a true understanding is formed of a process's structure that context-free concepts are acquired. A's new concept of a dipper will be transferable to other situations in which it might be useful, whereas B's notion is limited to the original sequence of leaf stripping followed by ant dipping.

It is interesting, too, that more complex combinations of the different levels of understanding are possible. The learner might grasp the significance of some bits of the procedure but not of others, so that at some levels she has a proper structural comprehension of what she is doing, but at others her actions are simply mindless repetitions. One of the implications of this fact is that not all cultural replication in modern human society will necessarily be memetic, just because the distinguishing feature of human culture is that some of it is. My argument is that the distinctively human form of culture, which may be called memetic, emerged on the back of the ability to meta-represent – but this does not entail that *all* elements of human culture are memetic. Humans are still capable of the mindless repetition of detailed actions, and of the almost equally mindless repetition of structured behaviours; it is just that we are also capable of much more.

The Beginnings of Memetic Evolution

It is this "much more" which explains why human culture has exploded in a way that no other organism has achieved. If your thoughts are firmly attached to your behaviour and environment then there is not a lot of scope for their expansion. If, however, you can free your representations from their external ties, then suddenly there is a whole new arena for their evolution.

Given the capacity to meta-represent the internal links between their representations, the mental activity of our ancestors would rapidly have been taken over by the new replicating concepts – just as in chemistry the primeval soup was soon taken over by the new physical replicators. In the physical case the replicators increased in numbers, but not in equal proportions: some chemical combinations had greater copying fidelity, fecundity or longevity, and were therefore more successful than others. The ultimately limited space, combined with this variation, led to competition between the different types of replicators – and this would also be likely to happen in the case of mental replicators. Some combinations of activities would have been more complex than others, and these may not have been transmitted as accurately as the simpler ones. Some activities would only have been appropriate in very specific circumstances, and these would necessarily have been less "fecund" than others: they would only have been transmitted to the few other creatures to whom they were useful. Still other activities may have been so brief, and so little different from their primitive components, that they did not have time to be transmitted before they were finished.

As with physical replicators, the environment would have had a significant effect on the success of the various cultural replicators. A peripatetic population, for example, would provide a more conducive environment for an inaccurate replicator than would a static population: in a never changing environment, a behavioural pattern that alters in transmission could soon become inappropriate for the surroundings, but in a shifting environment the pattern's mutation may well be more appropriate for the new surroundings than was the original.

Still, variation without limited resources does not make for selection. What restriction would there have been on the "resources" available to the new replicators as they became more numerous? There are a variety of answers to this question, some more controversial than others. Those who think that language is a prerequisite for thought might want to say that the initial emergence of thought would have been limited until language had developed to a certain level. Advocates of the theory that the brain expanded as a *result* of the emergence of culture would claim that initially the brain itself would impose an upper limit on the complexity and range of new ideas that could be grasped. It is certainly interesting to speculate about the positive feedback that may have occurred between hominid brain size and the tool use that was facilitated by abstract concepts: it has already been noted that carnivores tend to have larger brains than herbivores; once homo had learnt to conceptualize a primitive "blade", and thence to strike a sharp stone flake from a rock surface, he became able to acquire meat from even the toughest-skinned animals. It is tempting to infer that a positive feedback "loop" would thus have been set up between tool use and brain size.

Such conjectures are, however, hotly disputed by other scientists, and I should certainly not want to rest a hypothesis on them. Instead, it is simpler to say that the restrictions on the number of early cultural replicators would have stemmed from the fact that even in the most intelligent modern animal the brain's attention is limited, and no concept will succeed if it holds no interest for its recipients. This restriction on the number of new replicators that could persist, together with their variety, would have ensured that the replicators evolved.

Conclusions

This chapter has explored one version of the story of memetic origins. From the "primeval soup" of primitive mental activities, given the right stimulation, there emerged the first cultural replicators – the ancestors of memes. Modern memetic evolution could not begin, however, until our own ancestors were capable of manipulating – and of course copying – the right sort of mental representations. However memetic information is transmitted between individuals, the important thing is that it should be represented in such a form that those who acquire it can manipulate it freely, without being tied to a particular context. Many organisms can represent the world around them, but memes are representations with a particular nature: as complex, context-independent concepts, they depended for their emergence on the development of unique mental capacities. The participants in memetic evolution needed to be able to compare incoming information with their existing knowledge, to fit it in with their existing skills, and if necessary to rerepresent it in a different format – and the mere fact of its being transmitted via imitation cannot guarantee this. Meta-representation is the key.

Clearly this argument is founded on Chapter 3's contention that memes have their basis in representational content, but thus far little has been said about what *form* that content takes. How, in practice, is it realized? In the next chapter, I return to the search for cultural DNA.

11

Memetic DNA

In biology, what matters for evolution is that genes are able to *retain* (i.e., realize) information, which they can then *carry* to the next generation, where they then enable it to produce *results.* Genes are a method of storing information in such a way that it can be replicated and put to good use; the effects that it has must be possible in a variety of surroundings. To look at this another way, genes may be seen as representations, in DNA, of the phenotypic features that they control. All species use the same system of representation: DNA.

If we are to pursue the analogy with the biological world, then we should expect to find that what matters for cultural evolution is that memes are able to retain information, which they can carry to the next cultural generation, where they enable it to produce results. Memes are a method of storing information in such a way that it can be replicated and put to good use in a range of situations. To look at this another way, memes may be seen as representations of the phenotypic features that they control. The questions now arise what representational system (RS) they use, and in particular whether it is always the same system, as is the case in biology.

Words

I turn first to language which, in the picture of culture that has been painted so far, has been assigned no particular role. It would be worrying for meme theory if language really had no part to play in it, since language is one of the most significant elements in our cultural lives. A vigorous debate surrounds the question whether language is the medium or merely the communicator of thought, but few would deny that at least some thoughts are dependent on it. Putting this even less controversially, it is clear that language use fills a large part of most people's lives, and plays at least some role in their cognitive development. A theory of cultural evolution that misses out language will have quite a large hole in it.

Memes as Words?

It seems fortuitous, then, that languages – and words in particular – provide one of the most tempting answers to the question what form memes' RS takes. Words are certainly able to realize information, to carry it between people and to have an effect on them. Moreover, their apparent evolution has been noted by several commentators. An illustrative example is provided by the following passage from Richard Dawkins:[1]

> Languages clearly evolve in that they show trends, they diverge, and as the centuries go by after their divergence they become more and more mutually unintelligible. The numerous islands of the pacific provide a beautiful workshop for the study of language evolution. The languages of different islands clearly resemble each other, and their differences can be measured precisely by the numbers of words that differ between them, a measure that is closely analogous to the molecular taxonomic measures. . . . Difference between languages, measured in numbers of divergent words, can be plotted on a graph against distance between islands, measured in miles, and it turns out that the points on the graph fall on a curve whose precise mathematical shape tells us something about rates of diffusion from island to island. Words travelled by canoe, island-hopping at intervals proportional to the degree of remoteness of the islands concerned. Within any one island words change at a steady rate, in very much the same way as genes occasionally mutate. Any island, if completely isolated, would exhibit some evolutionary change in its language as time went by, and hence some divergence from the languages of other islands. Islands that are near each other obviously have a higher rate of word flow between them, via canoe, than islands that are far from each other. Their languages also have a more recent common ancestor than the languages of islands that are far apart. These phenomena, which explain the observed pattern of resemblances between near and distant islands, are closely analogous to the facts about finches on different islands of the Galápagos Archipelago which originally inspired Charles Darwin. Genes island-hop in the bodies of birds, just as words island-hop in canoes.

The mere fact that languages change over time is obviously not going to be enough to convince us that language should be given the same role in culture as DNA plays in biology, but here Dawkins notes an additional analogy. In biology it is possible to compare the same molecular sequences as they occur in different animals, to see *how* different they are: since each gene or protein has its own mutation rate, the molecular data – that is, the number of differences – can (given certain assumptions) provide a measure of how long it is since the separate species had a common ancestor. Dawkins points out the analogy between such molecular taxonomic measures for species and taxonomic measures of language in terms of word differences.

There's More to Language Than Words

There is, however, a potential weakness in this analogy. The reason why molecular taxonomy is appropriate in biology, is that DNA is something that all species have in *common*. The difference between the molecules of separate species' DNA is measured on the assumption that all of those species have not only *some* but the *same* genetic medium in common. Assuming that the analogy will work at all, then, it will certainly only work for different cultures that use a language which is common to all with the exception of a few words. Unfortunately, this is rarely the case. Although language *use* is obviously something that all cultures have in common, each individual language is not. It may be, therefore, that the differences between cultures' particular languages are so great as to render implausible the hypothesis that words form the basis of memes.

Apart from anything else it would be wrong to assume, on the basis of Dawkins's example, that language differences can always be appropriately measured by counting word differences. Although his example is very specifically about languages that differ only in this respect, as soon as broader cases are introduced (the difference between Latin and Italian, for example) the assumption breaks down. To argue otherwise would be a false extrapolation from the gene-meme analogy: although molecular differences might provide an appropriate measure of species differences, there is much more to take into consideration about languages than the words in their vocabulary. The combinations and ordering of those words, and the rules that govern them, are two significant factors which spring immediately to mind.[2]

Words and Their Meanings

Moreover, there seems often to be a worrying mismatch between the content and the linguistic expression of memes – a mismatch that appears, at first glance, to be incompatible with the suggestion that memetic content might be realized in words. Not only is there a variation amongst languages between cultures, but even within one small country like England there are different words for the same concept (think of interchangeable pairs such as: lorry and wagon; sofa and settee; pudding and dessert), and conversely the same word may have different meanings: terms such as bank, run and leaves are polysemous in this way. There are other words whose meaning, though fixed for the moment, has altered over time: an example is "artifice", which used to refer to craftsmanship but is now more commonly used to refer to cleverness or even deception. There are even words, known as contronyms, which are their own antonyms: an example is "cleave", which can mean both "adhere" and "separate".

Closer inspection reveals that not all of these cases pose a problem for the memes-as-words hypothesis. When it comes to interchangeable words, for instance, there would be no difficulty in claiming that memes sometimes have multiple realizations, giving rise to synonyms. Moreover, synonyms surely present a problem for the claim that there is a meme for every word, rather than for the hypothesis that there is a word for every meme.

Nonetheless, this does not eliminate the problem of ambiguity, which arises when a particular word has a variety of functions, depending on its context. This must be a genuine difficulty for the memes-as-words hypothesis, for it strongly implies that being a particular word does not fix the function and content of a meme. Of course we can use context to determine which role a word is playing: although two memes can be realized by the same word (e.g., bank), we should know *which* of the memes it is realizing at any given time, by taking into account the context. Indeed, it has been emphasized already that context *is* important to memes – but any adequate memetic medium must enable the content of a meme to be determined by the interaction between its underlying structure and its environment, rather than being fixed by its environment alone.

How Powerful Are Words?

Such a medium must also enable a meme to exercise control over the phenotypic effects that it represents – and words alone cannot do this. The mere fact that I understand a word is not enough to ensure the execution of the associated meme's phenotypic effects. The word "suicide" provides a clear illustration of this point (see Chapter 13 for further discussion of the suicide meme). A person may perfectly understand the meaning of that word, and may even be going through an extremely bad patch in his life, but these two facts in themselves would not necessarily be enough for him to think of killing himself. In other words, despite his understanding of the word, and despite the fact that the external circumstances are apparently conducive to his participating in the action described by the word, still he may not even consider so doing.

Of course, this emphasizes once more the importance of *memetic* context: he may not feel suicidal because of the nature of what he has accepted into the rest of his meme complex (e.g., he may believe the statements that things normally do get better as time goes on; that suicide leads inevitably to eternal separation from God; that even a bad life is better than death, etc.). Nonetheless, it also emphasizes the fact that the exercise of a meme's phenotypic effects depends on more than the possession in one's lexicon of a given word or phrase. In order for a meme to be able to exert its executive effect, its possessor must not only *understand* the concept involved: she must also *accept* that concept into her mental assembly. I understand perfectly well the concept of the tooth fairy, but I do not subscribe to the belief in such a being, which thus plays no executive role in the control of my behaviour.

Memes as Words – or Language?

The freedom that languages exercise in linking words to concepts; the part that human minds play in accepting or rejecting the concepts that words do carry; and the fact that words are not the sole constituents of natural language, all contribute to the rejection of the hypothesis that memes are realized in words per se.

It is possible to see the root of all these problems in the question what *is* a word? Is it a spelled unit or a phonetic unit? Is it individuated by lexical entry (so that at the deepest level there are no polysemous words – though still there would be natural language ambiguities)? Are there any words at all in linguistic systems such as sign language? What of other representational systems? Would the hypothesis that words realize memes help at all to resolve the issue whether memes are internally or externally realized? Questions such as these reveal the fact that words are too restrictive for the role of memetic realizers: we need to look further abroad and take into account the whole linguistic system of which they are elements.

Thus language may yet play a significant role in the realization of memes. Vocabulary is not the only significant factor for native speakers of any given language: they rely also on the internalisation of sets of standard formation rules. If we accept the importance of such rules (and can ignore for the moment the debate about their content), then instead of concentrating on particular natural languages and their vocabulary perhaps our search for the memetic medium should lead us next to examine *language use* at a more general level.

Language: A Representational System

In fact, I would go further and argue that language itself is too narrow to play the role of cultural DNA. Rather, the answer lies in our *general capacity for representation,* of which language is merely a particular (if ubiquitous) example. Natural languages are powerful systems of representation, but so are systems of musical notation, of mathematical symbolism, of sign languages,* of reading methods for the blind, and of secret coding. According to my theory, such systems can be seen as the languages of different cultures: of the musical and mathematical cultures, and so on. Whichever of these systems has the necessary properties (of realizing information, transmitting it and producing effects from it), that system is capable of realizing memes. The memetic analogue of DNA is, then, the capacity to represent in the stipulated way, or the use of such a system of representation. That is what underlies all cultures, and that is what this chapter sets out to explore.

The danger with this reply to the question of how memes are realized, is that it may seem to be too vague. Of course cultural change should be expected to have its basis in the aspects of culture that have some measure of permanence, and systems of representation are just those things that do provide durability for ideas, designs and tunes. But what is there in the suggestion that RSs are crucial to cultural evolution to imply that there also exists a distinct *unit* of cultural inheritance and selection, with characteristics such that it deserves to be compared to a gene? In order to answer this question it will be helpful first to discuss the nature of any RS that purports to fulfil this vital role in culture, before turning to the issue of whether its constitutive representations may realistically be characterized as "memes".

Representational Systems

Terence Deacon[3] draws our attention to three different ways of representing information. We can use *icons,* which resemble that which they represent: a road sign depicting a man digging or a car skidding is iconic, as is the picture of a printer on my word processor's toolbar. We can use *indices,* which are correlated with that which they represent: the position of the fuel gauge indicates the level of diesel in my car. Finally, we can use *symbols,* which represent via social convention or established code, rather than by resembling or being straightforwardly correlated with that which

* In fact it can be argued that at least some of these *are* natural languages.

they represent. An ichthus (fish) on someone's car symbolizes his Christian faith, and a treble clef symbolizes a particular convention about which notes are represented at each position on the musical stave.

The special feature of symbols, as opposed to icons or indices, is that their content depends on relationships between themselves, not just on correlations between each one and an object or idea. Because there is no straightforward relationship between a symbol and that which it represents, a symbolic representation depends for its meaning – hence for its effects – on its context within an overall RS. Only in virtue of their context within that system will changes in the sequence or make-up of words within a language, or in the order or position of notes on a stave, change the meaning and hence the phenotypic effects of the language or music involved. A series of black dots arranged on five parallel lines does not have any intrinsic meaning, and the fact that one of those dots is moved from the bottom to the top line has no meaning – hence no effect – outside the context of a particular system of musical notation.

There are two important consequences of this interdependence between symbols. The first is that a "logically complete system of relationships among the set of symbol tokens must be learned *before* the symbolic association between any one symbol token and an object can even be determined".[4] In other words, it is impossible to work out what a symbol represents, if you do not already understand the RS of which that symbol is part. The second consequence is that symbols can represent *each other*, as well as things in the world. For instance, the phrase "the fifth letter of the alphabet" is a string of symbols representing another symbol, "e". It is, in other words, a meta-representation.

Back to Genes

At first glance, when we consider cultural RSs like language or musical notation, it seems implausible that they should be analogous to DNA. One of the key features of DNA is that it is common to all species, but there are many different systems of representation in use in human cultures. How, then, can the two be examples of the same phenomenon?

The key to the answer to this question lies in the fact that, in the biological world, when an organism is created it has already been endowed with all the genes that it is ever going to have. Its parents have, in effect, given it representations of particular pieces of information, which will then have their characteristic effects on its ontogeny. In other words, organisms are born with a set of representations: an RS.

The meaning of each genetic representation is determined by the sequence and relative proportions of the four nucleotides which are the building blocks of DNA. None of the nucleotides has any intrinsic "meaning": indeed, three of them may also be found in RNA (ribonucleic acid) – a different system again – and ultimately, of course, all are composed of elements that may be found all over the physical world. The meaning of any particular sequence of nucleotides, or of any mutation in that sequence, is absolutely dependent on its context within the genetic system. Only mutations that occur within the limits of the DNA system (the inversion of a piece of chromosome, perhaps) will be meaningful and effective within that system. Were it possible for a mutation to occur which was meaningless within that system, then it would not be able to exercise any external control and might even result in the system's destruction.

Thus it is obvious that within organisms, which are endowed at conception with a particular RS, only the *content* of that RS can evolve. The same information will always be represented in the same way – *within* the given system. The copying method that is employed by the genetic system entails that genes need structural replication in order to replicate at all.

Memetic RSs

Minds are not created in this way: a mind can go on acquiring new information all through its life. The reason why it can do this, on my theory, is that it has the ability to represent: to acquire and then transmit ideas. A neonate organism *possesses* a set of (genetic) representations, but a neonate mind has the ability to *acquire* them. We are all susceptible to cultural input and, in addition to our ability to abstract information from the environment, given enough stimulus we are also able to learn how to represent that information in a powerful and replicable form.

Deacon[5] supports this view of humans as a symbolic species, and wonders what it might be about the mind of a child which opens the early learning window that is so crucial for language acquisition. His rather surprising response is to point up young children's poverty of short-term memory and concentration. He argues that this may actually be an advantage when it comes to language acquisition: it gives infants a head start by enabling them to see the structural skeleton of language's rules, before fleshing it out with details of individual word meanings. I would add that this coincides with what is needed to decode *any* RS: one needs first to look for the structural features beyond the details, before adding the details and complexity later. (Though notice that this obviously only works for creatures that *can* represent: poverty of memory and concentration clearly doesn't provide a language-learning advantage to guinea pigs or goldfish.)

Extending Deacon's idea to other areas of learning – since my own claim is that our representational abilities are not limited to language alone – it is interesting to note some of the nonlinguistic evidence in its favour. The claim that human infants focus on overarching structures rather than on fine details is supported by observations in nonlanguage areas such as the development of perceptual skills. From a very early age – often prelanguage – children can recognize a whole range of different objects as falling within one category. Having grasped the concept of "cat", for example (and even if they cannot yet say the word, indicating their recognition, instead, with a consistent noise like "iaow"), they will be able to pick out not only real cats but also soft toys, cartoons and the most sketchy of drawings or stylized ornaments as cats. Clearly their recognition revolves around certain key features which they have extracted from that type of object, rather than the finer details of each token cat, which often differ widely (a smooth marble ornament vs. furry soft toy, etc.). As Deacon puts it, we humans "cannot help but see the world in symbolic categorical terms, dividing it up according to opposed features, and organizing our lives according to themes and narratives".[6]

Given this innate capacity to represent – to abstract information from the environment and realize it in a different, more concrete form – a mind can develop. Its "ontogeny" will not be the result of an interaction between a given set of representations and the environment but rather a continuous process of interaction between its innate learning capacities and an environment that is filled with myriad representations, as well as much that can be represented.

Different RSs

Since the mind has this general ability to represent information (rather than just a particular set of representations), it is not restricted to the use of any one RS. Because it is able to abstract information not only from the environment but also from the representational form in which it is encountered, the mind can adopt new systems of representation as well as new representations within an existing system. This means that evolution can take place in both the content and the system of mental representations, making it not at all surprising that many different RSs have developed in culture, whereas only one is prevalent in biology. It is merely a consequence of the different natures of the two spheres.

What will determine the "choice" of RS for a particular area of culture? Clearly, as in any area of evolution, whatever develops will be limited by what went before, as well as by external context. There is an element of arbitrariness about the development of any area of evolution. In biology, for instance, there is an arbitrariness about the universal code that is used for constructing organisms from DNA: "there is no functional reason why a given codon should code for one amino acid rather than another."[7] (Indeed, this fact increases the plausibility of the hypothesis that all life has a common ancestor, since if the code *were* the only one functionally possible, then it would be used universally even if the organic world were *not* all interrelated.) We should not be surprised, then, to see in the variety amongst cultural RSs an element of arbitrariness rather than functional adaptation.

As has been emphasized previously, evolutionary theories provide only relative information, so we should not expect memetics to be able to tell us whether the RS used in any particular cultural area is the best for the job in absolute terms. Indeed, in some cases (such as the notation for differential calculus in mathematics) there may even be more than one RS available for the same job. Nonetheless we *should* expect each, in its specific area of culture, to display superiority over those that are used in other areas – and this is indeed apparent. The natural language RSs that have evolved for our usual, everyday needs, although facilitating simple, urgent communication ("Stop!"), also enable us to develop and communicate very complex thoughts. Alternatively, if the desired effect is musical then we use a much more concise RS, which can display all the information about a note's length, pitch and dynamics in one symbol – and this certainly serves its purpose better than natural languages would. The same can be said for RSs such as engineering drawings and mathematical symbolism.

Notice, too, that care needs to be taken in discussing the evolution of the various RSs, not to imply that we started with a desired effect and worked towards the best RS for its production. Rather, just as cultural content has evolved over the years, so has the form in which it is realized. Although in the case of biology it is only content that has evolved, and not the system in which it is realized, this does not show that biological and cultural RSs are not really instances of the same phenomenon. For both, what matters is that the information is realized in a form that facilitates its preservation, replication and empowerment: an appropriate RS. For neither does it matter which RS is involved, so long as it has the right properties. Despite their diversity, and the consequent need for translation between them, systems of representation like language, musical notation and numerals are the cultural equivalent of biology's RS, DNA.

Meta-Representation Again

The previous section attributed the evolution of different cultural RSs to the human ability to represent (as opposed to being endowed with a particular limited set of representations) – but of course this is not enough in itself. Plenty of species can do that, and there are few to which we should want to attribute any meaningful amount of culture. In reality, we are able to learn and develop new RSs, to compare them with each other and even to choose which we use on any given occasion, because we can *meta-represent*. It is this which enables us to abstract information from the representational form in which we encounter it. It is this, too, which accounts for the ways in which our noninnate concepts have increased so furiously, and our thought processes acquired such powerful machinery.

The Significance of Non-Linguistic RSs

As a starting point, it is clear that the very existence of different cultural RSs is enough to give a powerful boost to the numbers of noninnate concepts that are available to humans. Consider, for instance, the fact that nobody could have had a thought about the map of Spain until systems of mapmaking had been developed. As another example, before the thirteenth century, when musical notation of the form used today began to be developed, nobody could have had a thought about a stave, let alone have composed music using it. Such particular examples could more generally be expressed as the fact that nobody can have a thought about a *component* of an RS until the RS itself has been developed. This is because outside the context of a given system of representation the marks or conventions used within that system are meaningless: they carry no intrinsic meaning, but gain their significance from their position within the system. Thus the development of a new RS brings with it the potential for the emergence of a whole host of related concepts.

Another significant fact about the various cultural RSs is that they make it possible to represent the same information in different ways. At the most basic level, it is of course feasible to do so *within* a given RS: within the English language, for instance, we can use different words, such as "pair" and "brace", to represent the same idea. There is more scope, however, to represent the same information in different ways, if you begin to employ a variety of RSs: for example, "two" (English language),

"II" (Roman numeral), "2" (base 10), "10" (base 2). It is obvious that language holds the greatest potential for alternative ways of representing other RSs' representations – but it is important not to allow this fact to obscure the importance of those alternative, specialized RSs.

As an illustration, consider the expression "o'clock". Until the late seventeenth century, that term referred "to the sounding of the bell at the hour – all the Germanic, Latinate and Celtic cognates of English *clock* mean 'bell'".[8] Why? Because before that time, clocks were not in common use – and until they were, there was no RS that involved a round clock face with two hands representing the time of day. So not until the late seventeenth century could someone have had a thought about this symbol: 🕒. As soon as that thought *was* available, it was also expressible in a natural language: in the case of English, it would be expressed as "three o'clock", where "of the clock" now referred to a clock rather than to a bell's chime. Nonetheless, without the clock-face RS the linguistic version would have no meaning at all, as demonstrated by the fact that a French speaker could not acquire the "three o'clock" concept, even if I translated it into his language, unless he first acquired the appropriate RS. The words "il est trois heures" would have no meaning to him unless he understood the horological system upon which that phrase depends for its context.

The problem is that the ubiquitous use of language can make it harder to see the significance of the alternative, specialized RSs on which many of our concepts depend for their coherence. Natural languages have such a huge potential for representing other RSs' representations that they are capable of expressing almost every concept that originates in a specialist RS – and when we express such concepts linguistically, it can mask their dependence on the original RS. This in turn obscures the fact that it was the development of that RS which facilitated our capacity for those sorts of concept in the first place.

Yet this ability to move between RSs – to contemplate and select the way in which information is represented – is hugely important for memes. As is shown in the following section, it is only once we take into account nonlinguistic RSs like systems of mathematical or musical notation, the conventions of map making or horology, that the power of meta-representation really begins to emerge.

Comparing RSs

It is of course possible to meta-represent within natural languages: "Why should people find it so hard to spell 'Distin'?" is a sentence in which a question is asked, in English, about an English word. (Indeed in talking about that English sentence I am meta-representing it, also in English.) Not only this, but language is also the RS with the most potential for meta-representation: some systems, like musical notation, have limited potential for this level of representation, whereas in natural languages it is as easy to express a meta-representation ("the first letter of my surname") as a representation ("D"). The ubiquity of natural language is no accident; it is without a doubt the most powerful and adaptable RS that we humans have developed. My point is simply that it is when we begin to see meta-representation in the light of alternative, nonlinguistic RSs that its significance is most clearly illuminated.

One of the key features of meta-representational cognition is that it enables us to think about *how* we are representing a given piece of information. Indeed, it is often the *contrast* with an alternative system which will enable someone to see the crucial feature of his existing RS. In mathematics, for example, there is more than one way to represent a number: our traditional decimal system is one way, and the binary system that computers use is another. It is probable, however, that most people will not step outside our system and understand that place value is more important than the form of the individual numerals (each of which represents how many 1s, 10s, 100s, 1000s, and so on, are in a given number), until they are shown an alternative such as binary (where each numeral tells us how many 1s, 2s, 4s, 8s, etc., make up a number). Such examples demonstrate that we develop not only the content of our thoughts but also our capacity for thought when we begin to compare different RSs with each other: to meta-represent. Just as the previous chapter suggested, it is the human capacity for meta-representation which is the key to the evolution of our cognition and creativity.

Given this capacity, it is unsurprising that a variety of cultural RSs have emerged. Once you can think about how information is represented, it is inevitable that you should begin to develop alternative systems, seeking RSs that are more suited to whichever sort of information is involved. At the heart of human culture lies our ability to meta-represent. The cultural equivalent of DNA – that which all cultures have in common, which enables them to evolve and even to replicate – is not one, but a whole range of systems of representation.

Robert Aunger

I should note that Robert Aunger would strongly disagree with this claim. In his book *The Electric Meme*,[9] Aunger argues that replication must always be specific to one kind of physical substrate, since its aim is "to make sure only one kind of product comes out at the end".[10] He denies that "replicators can migrate from one form to another without consequence,"[11] pointing out that the influences which a replicator can wield are highly dependent on the medium in which it is realized. Information, on this view, cannot pass unscathed between media such as books, brains, audiotapes and computers: true replication involves "structural equivalence"[12] in the influences exerted by source and copy.

Clearly there is a fundamental discrepancy between Aunger's view of memes and mine: he believes that memetics would be scuppered by the claim that memes can successfully be represented in a variety of different media, whereas I would claim that the development of different RSs has been crucial to the evolution of human culture. Now, of course "representational system" cannot simplistically be equated with "medium". The same RS can be used in a variety of media (it is possible to use the English language in thought, writing, typing and speech), and conversely the same medium can realize a variety of RSs (it is possible to use pen and paper as a medium for the Hebrew language, for mathematical notation and for engineering drawings). Yet similar lines of reasoning can nonetheless be applied to both media and RSs, since it will often be the case that a change in one involves a change in the other. Thus Aunger's attack on the view that memes can be realized in a variety of media is also an attack on my sort of memetics – and the important points that he highlights in the course of this attack are as significant for my version of meme theory as for anyone else's.

Firstly, Aunger is right to emphasize that the medium in which information is stored has "a tremendous impact on the dynamics of evolution",[13] although he goes too far, I would argue, in inferring that replication must therefore always be "specific to one substrate".[14] The story of Little Red Riding Hood is a popular example amongst memeticists, and I do hold the line, contra Aunger, that it is possible to copy this tale from spoken French to written English, then to scan it from paper into a computer's memory, copy it thence to a CD, and so on. I would maintain that each of these versions really *does* represent the same information – but that the potential *effects* of that information will be facilitated or curtailed by the medium in which it is realized at any given time. It has already been noted that context will affect the results that stem from any replicator, and it is obvious that a crucial part of its context will be the medium in which the replicator is realized.

What does this mean in practice? At an obvious level, the spoken version would hold the attention of a monolingual French child in the way that the written English could not; but more fundamentally than this, each version plays a nontransferable role in a particular network of causal links. Aunger specifies that replicators should not be "defined as similar" unless they do "the same kind of job in the same kind of context" as each other,[15] and clearly the different versions of the tale do not fulfil his condition, since the contexts vary so widely. As a result of the medium in which it is realized, each version is preserved, accessed and experienced in quite different ways from all the others, and each exerts a unique range of influences. Yet I would challenge Aunger's assertion that this proves that each is not really a copy of the same information as all the others – that the same information cannot be realized in different media. The content of a meme is at least partly defined by the phenotypic effects that it controls, but it must never be forgotten that its content *is* independent of those effects: the fact that they may be implemented to different extents in different media does not mean that the content itself has been lost. Aunger is right to emphasize the importance of medium for evolutionary dynamics, but wrong to infer that replication cannot, therefore, take place across media. Changes in memetic medium are not insignificant – but neither are they impossible.

The second important point to come out of Aunger's analysis is that choice of medium affects not only the potential influence of memetic information but also its content. In maintaining that it is possible to preserve the same information in a variety of media, I do not wish to turn a blind eye to the difficulties that are intrinsic in such a process. Choice of medium will inevitably limit information's content, just as choice of RS does: there is a vast difference in quality of sound between audiotape, CD and vinyl music recordings, and the English adopt words like "schadenfreude" precisely because there is no equivalent in our language.

This means that information may well be lost in the course of changes between media, just as it may be in translations between RSs. What it does not mean, however, is that the essential elements of that information may not legitimately be considered to have been realized across the range of media and RSs in which it has been represented. Just because a CD represents an analogue sound wave via a stream of digital numbers (different RS *and* different medium), we do not say that it is not really a recording of the music which those numbers represent – although we may well say that it does not sound as good as it might have done, had it been recorded on a good vinyl record.

Nor does it mean that information will *necessarily* be lost in the course of media change. If I want a bit of technical detail to flesh out a philosophical example, then I might e-mail a question to my husband, in the hope that he may mull it over in his lunch hour at work. His response will first be formulated in his brain, thence transferred to type on a computer screen before being transmitted electronically to my own computer, possibly printed out on to paper, and finally read by me. If he has expressed himself clearly enough, then the information that is represented in my brain at the end of this process will be *the same* information as was represented in his at the beginning of it. The ways in which I can use that information, my understanding and application of it will all serve to verify this fact: it will be identical to his in content and effects, unaltered by its journey through brain, computer and page. To put it even more simply: Calum remembers the first line of a poem; Calum writes it down; Luke reads it and memorizes it; Luke writes it down.... In what sense is the replication in this sequence undermined by the media changes that it involves?

Nonetheless, it would be foolish to deny the significance of medium for representation – and indeed this has been highlighted by my hypothesis that in different areas of culture different RSs have evolved, the better to facilitate representation in each arena. I would argue that the same is true of the development of different cultural media. In both cases, the development of a range of representational options supports the view that the choice amongst them is significant, and that movement between them may be risky. Nobody would deny this. What it does not support, however, is the view that choice amongst them is illusory and movement between them impossible. The ultimate test, of course, would be for the final move in any chain of replication to be a return to the original medium: is the resultant representation identical to the original? If so, then it seems to me unproblematic to call the process by which that point was reached "replication".

Thus I can still claim that memes can be preserved and copied in a variety of cultural RSs (as well as across a range of different media). What matters for evolution is that the RSs have the right properties. Amongst those properties is one that is highly significant but has not yet been examined in any close detail, which is that the representations within each must be particulate. This is crucial, because if not, then individual cultural representations cannot realistically be characterized as "memes".

Particulate Memes

It was the great achievement of Gregor Mendel to show that hereditary units can be treated in practice as indivisible and independent particles. Nowadays we know that this is a little too simple. Even a cistron is occasionally divisible and any two genes on the same chromosome are not wholly independent. What I [Dawkins] have done is to define a gene as a unit which, to a high degree, *approaches* the ideal

> of indivisible particulateness. A gene is not indivisible, but it is seldom divided. It is either definitely present or definitely absent in the body of any given individual. A gene travels intact from grandparent to grandchild, passing straight through the intermediate generation without being merged with other genes. If genes continually blended with each other, natural selection as we now understand it would be impossible.[16]

Is there any good reason for supposing that cultural representations, too, might be distinct units of selection, playing the role in culture that genes fulfil in nature? Fortunately for memetics, the answer is "yes". In fact, there is just the same evidence in favour of this supposition as Mendel had for conjecturing the existence of genes: observation of their phenotypic effects.

Mendel's famous experiments on peas demonstrated that the characteristics inherited by one generation from the previous one are not blended with each other but may be regarded as indivisible. As Dawkins points out, any given gene is either present or absent in any given organism. There is nothing in between. The evidence for its presence or absence may be provided, today, by molecular analysis. Alternatively, it may be demonstrated as it originally was by Mendel, by observing the presence or absence of its phenotypic effects (though this observation may take several generations, e.g., in the case of recessive genes). Although knowledge of the brain is not yet advanced enough for neural examination to reveal the presence or absence of any given idea or skill, the external observation of its effects is an alternative that is certainly available in culture, as it was in biology. What does such observation tell us?

It tells us that there do indeed exist units of information whose "all or nothing" presence or absence may be discerned in the mind of an individual. This is not to say that a particular skill or idea will be manifest in exactly the same way in everyone who possesses it, for memes' phenotypic effects, like genes', will vary according to their environment and context. It is rather to claim that there exists, as Dawkins puts it, "an essential basis of the idea which is held in common by all brains that understand [it] . . . The *differences* in the ways that people represent [it] . . . are then, by definition, not part of [it]."[17]

What Does Common Sense Dictate?

Yet this discreteness has been challenged by many of memetics' critics. Maurice Bloch, for example, has asserted that "In reality, culture simply does not normally divide up into naturally discernible bits."[18] He says that it is almost impossible to specify what elements form each meme, and that it does not help to introduce the concept of meme complexes, because memeticists are "no more able to establish boundaries around these memeplexes than around the constituent memes."

In opposition to this claim, I would argue that "in reality" we can and do talk about people having skill *x*, understanding theory *y* or knowing tune *z*, meaning by this that they are in possession of a particular amount and type of information, without which they would not have *x*, *y* or *z*. They are able to pass on *x*, *y* or *z* to other people, and to do so in such a way that, even if the recipient manifests it differently, she too will have a copy of *x*, *y* or *z* that she can pass on to others. In each recipient it runs the risk of embellishment, corruption or diminishment, but this is no different from the risk that each gene runs of mutation during replication. The important point, though, is that we do not acquire information and skills in an indistinct, amalgamated form: we acquire discernibly separate units that are individually available for discrete retransmission.

Perhaps the best example of this process in action can be seen in the existence of dictionaries, whose very purpose is to collate and store concepts' content in a form that displays their essential basis. By defining a concept, the dictionary tells us what information we will know once we have acquired it; it also allows us to see the mistakes, omissions and augmentations in our own versions of it; and as dictionaries are updated we can trace the concept's evolution, when some of the variations come into dominance. Of course dictionaries are not the only stores of memes: they merely provide a helpful illustration of the discreteness of cultural representations.

Information and Actions

But Bloch has a more theoretical reason why this discreteness is implausible. He says that cultural knowledge is largely "inseparable from action":[19] the information that it carries is only partly "intellectual in character", and should not be characterized as a "library of propositions" which can be transmitted in discrete units.

Fortunately, as with many of the criticisms that have been considered throughout this book, it is possible both to accept this viewpoint and to continue to support meme theory. From the perspective of memetics, it is unsurprising that cultural information should be inseparable from action, for an integral part of memes is their executive role in producing phenotypic effects. Nevertheless, the information itself can still be transmitted in discrete chunks, even if there is some superficial blending between the actions that stem from it.

For example, someone who is able both to read music and to play the violin may appear at first sight to have blended these skills into one complex action when she does both simultaneously. It becomes clear that she has a set of separate, discrete skills, however, when she moves between different contexts: in church she can follow the music when singing hymns (even if her voice has not been trained to produce the notes accurately); given a mandolin she could both read the music and match the information to the appropriate positions of her fingers on the instrument (although her right hand would struggle); given a viola she could both read the music and handle the instrument and bow appropriately (although her left hand would struggle). Thus, although her skills are apparently blended when she plays the violin, still they can be used separately in other contexts. The close links between information and action need not, then, present a problem for the thesis that information is transmitted in discrete units.

Translation Versus Transformation

Yet Bloch wields a further argument against the particulateness of memes. Echoing Sperber's claim that straightforward copying cannot account for meme transmission, Bloch contends that any novel trait which an individual accepts must inevitably be modified so as to be coherent within its new context. In the transmission of culture, "Nothing is passed on; rather, a communication link is established which then requires an act of *re-creation* on the part of the receiver."[20] Copying cultural traits relies on "active psychological processes occurring in people"[21] – not "transmission between passive receptors" – and the resultant act of re-creation totally transforms the "original stimulus and integrates it into a different mental universe so that it loses its identity and specificity". How, whilst undergoing such a process of transformation, could memes maintain their particulateness?

There are two separate issues here. First there is the question how active our minds are when engaged in copying and retaining cultural information – an issue which has been raised several times already, and which will be addressed fully in Chapter 12. Secondly, there is the claim that information is so altered by the re-creation that is inevitably involved in cultural transmission that it cannot realistically be characterized as unitary – and there are several reasons why this view is unnecessarily pessimistic.

One reason is that, "contrary to what we were all taught in high school, genes are nothing like beads on a string. So both memes and genes are likely to have comparably complex structures."[22] Against this background it is unsurprising to find that memes sometimes *appear* to have blended – just as it would be easy to suppose, fallaciously, that the skin colour of a mixed-race child is the result of genetic blending. Just as in the child's case appearances are deceiving, so the appearance of memetic blending proves nothing very much. Human communication may involve some translation and interpretation, but this does not entail that it inevitably involves radical transformation.

As noted above in the example of the violinist, context is of course important in determining the behaviour that will result from a particular skill or piece of knowledge, and in some contexts it will not be possible for any results to be effected (e.g., when the violinist is driving her car). Bloch has pointed up the additional fact that context is just as important for *transmission* as it is for the production of behavioural effects, and no one could disagree with that: all good teachers know that information is best communicated when related to pupils' existing knowledge and life experiences. Putting this another way, information must be represented in a common language before it can be transmitted. But none of this entails that translation must always involve transformation. Indeed the very point of a good translation is to represent the *same* information in a different system. Similarly, its integration into a "different mental universe" will not inevitably alter the essential character of a portion of information. There is a real sense in which the violinist has the *same* skill as her violin teacher, despite their very different personalities and backgrounds. There is a real sense, too, in which memes can defensibly be regarded as discrete units of cultural selection.

Chomsky's Language Acquisition Device

My contention remains, then, that cultural evolution is based in a variety of representational systems. I have answered the objections that the elements of cultural RSs are not particulate, and that replication cannot possibly take place across media boundaries, but a pressing concern remains: is language really no more significant than any other cultural RS? The hugely influential American linguist Noam Chomsky has famously put forward the theory that language – or at least a "language acquisition device" – is innate: it is uniquely significant for humans, and has evolved along certain rigidly defined pathways. Is Chomsky's picture of human mental development an obstacle to my intention to bring a huge variety of RSs under the umbrella of "cultural languages"?

The Uniformity of Language

Chomsky's thesis is based on observations of the ways in which children acquire language, and in particular on the fact that infants within a linguistic community are not explicitly taught language, but pick it up from an extremely impoverished input whose content will vary enormously between individuals. Despite the apparent *inadequacy* of the input, children acquire intricate linguistic abilities with remarkable speed. Despite the *variation* amongst the input, their resulting language is (in all important respects) uniform. The only plausible explanation of these facts, say Chomsky and his followers, is that our early language acquisition is instinctive.

This innateness explains not only the speed and ease with which we pick up language, but also the uniformity: "Scope and limits are intimately related . . . the fact that there are many imaginable languages that we could not develop through the exercise of the language faculty is a consequence of the innate endowment that made it possible for us to attain our knowledge of English or some other human language."[23] In other words, given limited English input, our language instinct will enable us easily and swiftly to produce English output, and only English output. In providing us with a way to systematize the input, it also *limits* us to that way.

A further consequence of our being so limited is that there will be no fundamental differences between any of the natural languages. Experiential differences will of course lead to minor variations, but it is only a matter of degree between the cross-dialect and the cross-lingual differences, not a matter of type. This is comparable to the differences that will arise in your genetically determined body size, depending on the nutrition and exercise you receive as an infant.

Nonuniform Nonlinguistic RSs

Given this claim about the essential similarity between all natural languages, how much of a problem is raised by Chomsky's language instinct hypothesis for my insistence that we take into account *non*-linguistic RSs? It must raise some difficulties, for clearly there *are* fundamental differences between culture's nonlinguistic RSs, never mind between them and language.

The different numerical systems within mathematics again provide a good example here. Consider, for instance, the medieval Arabs' introduction of zero into the counting system: nobody could deny that this was a significant departure from all that had gone before; indeed zero has been described as "possibly the most significant mathematical figure of all".[24] Moreover, in the Arabic counting system the forms of the individual cardinals (0, 1, 2, ...) are less important than the place-value convention on which it is based – a convention which in itself is, without doubt, a genuine novelty in the method of representing numbers. It enables ten symbols to be used to represent an infinity of numbers, and is the reason why the Arabic system is so much more convenient than Roman numerals – an unwieldy system that depended on the form of the numerals involved (I, V, X, ...) and the rules for replacing one with another. The place-value convention really comes into its own in the conversion between different bases, to which I have alluded, which relies on the fundamental concept that successive digits represent successive powers of the base. In the binary system, for instance, there are only two alternative symbols (1 and 0), and their position is *all* that determines which number is represented by a combination of them.

It is clear, then, that there *are* genuine differences between RSs, even when they are used to represent the same cultural (in this case mathematical) information: the Roman and Arabic systems differ not only in the most obvious surface features (the number eight looks like this – 8 – in one system, and like this – VIII – in the other), but also in their most fundamental, underlying conventions. If we are permitted to characterize these nonverbal RSs as cultural "languages", then how does this square with Chomsky's theory? He says that "the human language faculty will never grow anything but one of the possible human languages, a narrowly constrained set," where the constraints are due to that faculty's "roughly comparable rule systems of highly specific structure".[25] Yet if we take into account the many nonverbal cultural RSs which play varyingly important parts in our communicative and cognitive lives, then we can see a vast diversity of rules and conventions displayed across them.

The Meta-Representation Device

As I have emphasized, I do not wish to underplay the significance of language. Chomsky's language instinct is now widely (though not universally) accepted, and brings a host of linguistic facts under one explanatory umbrella. What I should like to do, then, is not to reject Chomsky's theory but to broaden its implications. In summary, my own hypothesis is that natural language is so important for humans that an "instinct" for its development did indeed evolve – but that with that instinct came an equally significant ability: the capacity for meta-representation. The irony here was that, although this capacity had evolved as part of the language instinct, its emergence freed humans to represent in all sorts of nonlinguistic ways, thereby to weave a complex cultural web of mental and material artefacts.

Clearly, this hypothesis needs to be unpacked and defended. As a starting point, I should reemphasize the fact that I am persuaded by Chomsky both that there is an innate element to language acquisition and about the primacy of language amongst human RSs. Nor would I wish to challenge his claim that natural languages are constrained by a universal grammar, "the biological endowment that determines the general structure of the language faculty".[26] Human infants have an innate and highly specific "rule system", which gives them advance knowledge of the sort of grammar that will govern whichever language they experience. This has the double advantage of boosting children's ease and speed of language acquisition, and of ensuring that human languages develop along lines that are universally comprehensible.

So how can the many nonlinguistic cultural RSs – whose rules and structures are so very diverse – spring from the same faculty that gives us natural language? The answer lies in the fact that there is more to language acquisition than the universal grammar. In order to learn a language humans need a whole range of mental abilities. We must be able, for a start, to represent: to abstract information from the environment, and to realize that information in a manipulable, memorable and widely applicable format. We also need to be able to abstract information from an incoming RS, so that exposure to a sample of representations enables us to acquire the RS of which they are tokens. In order to do this, we need to be able to compare our representations and abstract their common features: to meta-represent. Clearly it would help if we not only are *able* to represent and to compare representations, but also *tend* to do this. It would also help if we tend to assume that incoming representations are subject to rules.

The plasticity of the infant brain is important, too. The more causal connections are laid down in the brain, the more they constrain subsequent reaction to and acquisition of novel representations and RSs, especially given our hypothesized tendency to compare our representations. As an infant, the brain is free from such constraints and open to influence: children need to be especially receptive, for they have so much to learn in such a short time. Moreover, as Deacon has pointed out, the mental capacities of young children have some features that appear to put them at a disadvantage – an underdeveloped short-term memory and poor concentration skills – but which may actually be an advantage in language acquisition. Rather than becoming bogged down in the complexity of the language around them, infants of necessity look first at the structural skeleton of linguistic rules, fleshing out the details later as their brains mature – and this is just what is needed for the effective decoding of any novel RS, since representations do not make sense without the RS within which they are embedded.

The latter fact also helps to ensure that children's language acquisition is not only swift but in addition remarkably uniform compared with their varied linguistic input. When children try to make sense of the language with which they are constantly bombarded (either directly or overheard), there are tight constraints on the way in which they can succeed: English input does not make sense unless you work out the rules that govern it. Of course this does not have to be a conscious process (compare the fact that I can play any harmonic minor scale you ask of me, on the piano, with much more speed than I would be able to verbalize the series of intervals that govern such scales), but nonetheless if humans have an innate tendency to compare incoming representations, and to assume that they are rule-governed, then children's success in coming up with the *correct* rules for their native language is unsurprising. So long as we tend to compare our representations, and to look for commonalities amongst them, then we will tend to discover the commonalities not only in their content (i.e., their meaning) but also in the rules that govern them.

When it comes to linguistic RSs, human infants have an additional advantage in their innate knowledge of the rules of the universal grammar. It would be impossible for us to learn language without the assistance of this biologically determined structure. Crucially, however, the preceding paragraphs have emphasized the fact that it would be equally impossible if we could not meta-represent. This matters because once we *could* meta-represent we could use that ability to develop other, nonlinguistic RSs: as soon as you can think and talk about representations (rather than just about that which is represented), then you can begin to change their RS. You can use your capacity to meta-represent in order to think about your current methods of representation and how they might be altered and improved; you can use it to choose how to represent novel information; it can help you to interpret and learn a novel RS. In short, the "language instinct" package contains a variety of mental abilities, including one (meta-representation) that also facilitates the emergence of nonlinguistic RSs.

Yet how, if these alternative RSs are supported by the abilities that comprise our language acquisition device, is it possible that they are not subject to its structural rules? Chomsky points out that ability and constraint are two sides of the same coin: the fact that it is the universal grammar which *enables* us to acquire language means, conversely, that we can *only* acquire languages that conform to that biologically endowed structure; "there are many imaginable languages that we could not develop through the exercise of the language faculty".[27] Again, I have no quarrel with this claim – but the price of liberation is paid for nonlinguistic RSs in a different currency: one in which *lack of ability* and *lack of constraint* are two sides of the same coin. The fact that it is *not* the universal grammar which enables us to acquire the alternative RSs means, conversely, that those systems need *not* be subject to its constraints.

It is, however, interesting to note that the universal grammar appears to exercise a dictator's control over every *oral* RS: every natural language. We should always beware of equating limitations in our imagination with limitations in reality, but I cannot think of a single alternative RS which does not depend in some way on the support of material artefacts: mathematical and musical notation are written systems; systems of horology appear on artefacts like clock faces; semaphore needs flags; and so on. It seems that, once our language has been acquired with the help of universal grammar, it can be used as a meta-representational system for the development of unconstrained alternative RSs – but that those other RSs will need additional, physical support to get off the ground.

Language, on this view, is so important that we have evolved an innate structure for its acquisition, a side effect of which has been the ability to develop alternative, nonlinguistic RSs. Yet those alternative RSs cannot be realized in speech, which has been developed for the communication of languages which follow the rules of universal grammar. RSs that are independent of those rules cannot be supported by a system which relies on the rules for its coherence. This is true even though we can of course find ways to express nonlinguistic representations in language: "starting on middle C, play the notes C, C, G, G, A, A, G," for instance, or "multiply two by three, and then add four." Although our meta-representational capacity allows us to translate information from nonlinguistic RSs into language, as in these examples, it is a struggle to communicate that information if we are restricted to speech alone.

In other words, we can usually find ways to *represent* the information in language, but it is harder to *realize* those representations in the primary *medium* of language (speech). All representations need a medium in which they can be preserved and manipulated, but my suggestion is that our minds struggle to manipulate nonlinguistic representations in the medium of speech – and that this is because language use is constrained by the rules of the universal grammar. The language acquisition device may contain an element that enables us to develop nonlinguistic RSs, but it will not then support the realization of those RSs in its native medium. That is why we rely on artefactual assistance (such as pen and paper) for their realization and manipulation.

Conclusions

This chapter has argued that the memetic equivalent of DNA is not one, but many cultural systems of representation. Unlike organisms and their DNA, we are not endowed at birth with one fixed RS, but have the capacity to learn and develop many varied systems. Language has primacy amongst them in that it alone is the result of a biological endowment which also facilitates its communication through speech. Crucially, however, it is also the result of the human capacity for meta-representation, and it is this which facilitates the development of alternative RSs. These nonlinguistic systems, whose rules and structures are incredibly diverse, must be realized in a medium which is not subject to the constraints of universal grammar.

I have defended this hypothesis against the charges that true replication is not possible across media, and that cultural information is not genuinely particulate. Conversely, I have suggested that it is supported by its compatibility with the most widely accepted theory of language, Chomsky's language acquisition device.

There have been echoes, throughout these discussions, of a now familiar issue. If the development of cultural RSs depends on our biological endowment (e.g., our innate knowledge of language rules and ability to meta-represent), so that our acquisition of memes is tightly interwoven with the development of our minds, then what is the relationship between the two? How independent are memes of the mind? The next chapter asks to what extent cultural evolution is driven by our own mental faculties, rather than by the memes which are its units of replication.

12

Memes and the Mind

It is time to explore in more detail the relation between memes and the mind. In the case of genes and the body, the relationship is one between a survival machine and the replicators that are its formative constituents. This is a reciprocal relationship, in which the body is built (and in some ways acts) in accordance with a genetic blueprint, and the genes are selected via their phenotypic effects, which in combination produce an individual organism. To what extent is the relationship between memes and the mind an analogous one? Are memes self-replicators, or are they more like passive pieces of information, wholly dependent on human minds for their activation – much as genes depend on the cellular apparatus to make copies of themselves?

The nature of the memes-mind relationship has been a recurring issue throughout the discussions so far. The Dennett-Blackmore hypothesis is that there is in reality no distinction between the two. An alternative view is that a significant part of our mental architecture is determined by our genotype, with cultural input making only a superficial impact on our mental capacities. My own thesis has been that our innate (i.e., endowed by our genes) mental potential is developed by interacting with our environment – a crucial element of which is memetic. This is not to deny the novelty and autonomy of cultural evolution as a genuinely different process from Darwinian selection in the natural world; it is simply to acknowledge that the mind's evolution is ultimately dependent on its genetic roots.

Beliefs as Memes?

The main threat to my proposition comes from theorists such as Dennett and Blackmore, who argue that the self is a vast complex of memes: humans should be seen as "the clever imitation machine taking part in this new evolutionary process, rather than a conscious entity who can stand outside of it and direct it".[1] That this thesis sounds bizarre and unappealing is not sufficient grounds for its dismissal. Indeed, one of the reasons why we find such an idea hard to accept, says Blackmore, is that memes are incredibly good at deceiving us (of course she doesn't mean that they do this consciously): they can gain a huge advantage by becoming closely associated with our idea of "self". She asks us to imagine two memes, one which represents an idea and the other a belief in that idea: in the memetic struggle for survival, she suggests, the belief is bound to be selected over the idea. Beliefs will gain the advantage because we tend to defend them and try to persuade others to share them, whilst at the same time – by being expressed as "my belief" – they encourage our conviction "that there is a real self at the centre of it all."[2]

There are several reasons why I am not persuaded by this particular belief of Blackmore's. Firstly, beliefs are not memes, but *responses to* memes. Even when someone is doing her best to persuade you to share her beliefs, the most she can do is to present information to you in a format which she hopes will encourage you to adopt her own approach to it. We are all familiar with at least some of the beliefs that our friends hold dear, on subjects like politics, religion and child rearing, yet each of us holds a variety of attitudes to those beliefs: some we also hold dear, some we reject totally, and on others we retain an open mind.

A term from philosophy, "propositional attitudes", is illuminating here. Given a proposition, such as "it will rain today," I can hold a range of attitudes towards it: I can *hope that* it will rain today, or *believe that* it will rain today, or be in one of any number of mental states in relation to that proposition; those mental states are my "propositional attitudes". Now, this is a concept shrouded in some philosophical controversy, which I could not hope to disperse here – but nonetheless it emphasises the fact that we react in various ways to the information with which we are presented. Belief is simply one of those possible reactions.

Moreover, on closer inspection it is clear that Blackmore's reference to beliefs as memes actually begs the question in favour of her main hypothesis: only *if* (as she claims) the self is a meme-complex can mental states such as beliefs and desires count as memes. If, on the other hand, there is a genuine distinction between memes and the minds with which they interact, then beliefs will more accurately be seen as mental states than as the information to which those states pertain. The self, on this view, is a conscious entity which responds to incoming information in a variety of ways, both cognitively and emotionally. Cultural information, then, is something separate from the agents who process it. If this is the case, then belief is one of the ways in which agents can respond to that information, rather than another piece of information to be replicated.

Blackmore can only contradict this (categorising beliefs as memes) if she has already demonstrated that there is no real distinction between minds and the information that they process – that our minds are simply a conglomerate of absorbed cultural information, and consciousness an illusion. Such a claim is of course hardly uncontroversial: many people would find it hard to be convinced by any theory of mind that has as a result her contention that "there is no 'I' who 'holds' the opinions."[3]

The Mind as a Muscle

In contrast, Rosaria Conte claims that the view of the mind as meme complex "arises from an insufficient understanding of the autonomy of (memetic) agents".[4] For Conte, "replication is the responsibility of the memetic agent," and "memes do not have to be clever; rather, meme receivers or interpreters do."[5]

Yet this still leaves open the question of how memes interact with their receivers. If the mind is neither created by memes nor simply a complex of memes – if a mind is something that possesses, rather than being composed of, concepts – then how does it develop the "cleverness" that it needs to deal with cultural complexities? There is certainly a sense in which minds *are* concept-dependent, in that they will not fully develop until they acquire some concepts.

In order to reconcile these facts, it is helpful to notice that if a mind cannot develop without acquiring some concepts, and yet it consistently does so develop, then there must be an innate ability to acquire concepts. In other words, there must be some innate mentality *before* the concepts are acquired, as discussed in previous chapters. If this is the case, then memes are not formative constituents of the mind in the same way that genes build the body, but rather are part of the environment that allows the mind to develop.

This suggestion can be illustrated by analogy with the development of a muscle. Infants have the basis of and potential for strong muscles, but in a form that is by no means fully developed. Once a muscle begins to be used, however, it soon strengthens and develops its potential. In this sense, exercise "creates" the strong muscle; but conversely the exercise could not have occurred in the first place without the existing basis of a weaker muscle.

Similarly, in the mental activity of a newborn child there is the basis of and potential for a fully fledged mind. As soon as this is put to use and begins to acquire concepts (both from its contemporaries and as a result of its own discoveries about its surroundings), it begins to develop that potential. Thus the concepts that it acquires "create" the mind only in the sense that exercise "creates" muscles: the mind itself does not merely *consist* of a complex of concepts, but rather develops as a result of its interaction with them. Furthermore, the concepts themselves would not have existed in the first place if there were no prior existence of some mental activity.

The difference, of course, between a muscle and the mind is that in the case of a muscle the only exercise that can strengthen it is that which stems from itself. The mind, on the other hand, may be developed by concepts that spring from sources external to itself: from other minds. This is due to the nature of memes as cultural replicators, transmissible between different people's minds in a way that exercise is obviously not transmissible between different people's muscles. Nor do the physical muscles develop as a result of instructions delivered by anything outside the body of which they are part, whereas mental development happens as a result of the executive power of the replicators acquired.

The Parable of the Sower

Yet this distinction between instructions and effects touches on another unresolved issue: to what extent do we have choices and control over our responses to incoming information and experiences? In rejecting the view of the mind as a meme complex, I do not mean to sweep under the carpet the question of how much control we really have over our reactions to novel memes.

A concept from counselling may be helpful in formulating a response. Some therapists talk about the "scripts" that we are given by our families, which are the messages that we receive from those around us when we are very young, and which we each interpret in different ways. They influence our beliefs about the sorts of people we are, about the ways in which it is normal to behave, and about what "life plans" we ought to follow. The problem is that "although the life script we write so early is highly influential, it remains largely outside our awareness."[6] In response to this, good counsellors try to decrease the "scripted" elements of their own reactions, by working to expand the parts of themselves of which they are aware and over which they consequently have more control; in this way they are more able to help their clients to do the same.

For example, if someone was brought up in a household where negative emotions like fear or anger were suppressed, then her script may well include the unconscious assumption that it is wrong to be open about such feelings, and her responses to a person who does display these feelings may include an automatic attempt to soothe them away. Unless she becomes aware of this assumption, she will not be able to make choices about whether to retain or reject it. Once it is uncovered, she can then choose to modify her responses, and rather than trying to avoid negative emotions she may feel more able to acknowledge and cope with others' strong feelings.

Clearly our responses will sometimes be "scripted" by our innate personality traits as well as by early environmental influences, and this raises the possibility that the boundary between the innate and cultural aspects of our scripts may not be as clear-cut as it at first appears. It raises the possibility, too, that our cultural influences are more disparate than I have so far acknowledged. In particular, there may be a real difference between two sorts of cultural input. Some will be part of our "script", and thus fulfil a role rather like that which Dennett and Blackmore envisage: the information that we have absorbed in the past will form the basis for our reactions to that which we encounter in the future, and there will be no clear distinction between our "selves" and the memes that we have so deeply absorbed. There will, on the other hand, be other cultural input over which we *do* make conscious choices, of which we *are* aware, and which we *do* deal with more "actively" – although the fact that counsellors and their clients are often engaged in work to uncover and transform the "scripted" elements of their behavioural responses suggests that this distinction may itself be rather hazy.

Still, this discussion highlights the fact that it does make sense to talk about the distinction between "us" and "memes": just because we cannot always accurately distinguish our "selves" from the mouldings of our backgrounds and experiences, it does not mean that there is no such distinction to be made. Although some of our responses are unconsciously directed by memes absorbed in the past, we do have a large amount of control over our responses to much incoming information.

On this view, consciousness cannot be explained as a meme machine, but rather the memes-mind relationship was more accurately portrayed two thousand years ago, in the parable of the sower.[7] In that parable, a farmer sows seeds in a variety of soils, with differing outcomes: the seed on the path is quickly eaten by birds; the plants that grow from seed sown on rocky places are soon scorched by the sun; the plants that grow amongst thorns are soon choked; but the seed that falls on good soil grows strongly, multiplying many times over. The message is clear: different people (and even the same person at different times and stages of her life) will respond to the same information in very different ways. Incoming information – the seeds of the parable – will be understood, remembered, acted upon and then passed on to others with varying degrees of accuracy and enthusiasm, depending on its recipient's mind – the soil.

Of course the "type of soil" will be determined to a certain extent by the recipient's innate personality, and by his past experiences and cultural background. The potential within the "seeds" will be realized in different ways, depending on both his genotype and his current memotype. Yet the influences that these exert will not wholly determine the outcome. Unlike the much stronger claim that the mind *is* a meme complex, my thesis is that the choices we make about incoming data will be influenced by our existing memes – not that those choices are illusory.

Directed Evolution?

Yet the question remains how two apparently incompatible claims can be reconciled: on the one hand I am claiming that the mind *is* conscious, and our sense of self is based on reality, but on the other hand I support the theory that the cultural realm develops via an *un*conscious evolutionary algorithm. This echoes a point raised in Chapter 5, about whether the direction of memetic variation has (like the direction of genetic variation) no bias towards increased fitness, or whether it is directed by intentional human decisions.

Dennett and Blackmore reconcile these two claims by the simple expedient of rejecting one of them: they deny that the mind really has intentionality and consciousness. An alternative response is to claim that the direction of memetic variation is *both* unbiased towards increased fitness, *and* directed by intentional human decisions. How can this be possible? In the following sections I use an example from the development of engineering designs to demonstrate how two levels of description – intentional and mindless – can apply to the same process, and thus how the two can be reconciled. Consciousness does not have to be characterized as a meme complex for it to be plausible that memes participate in a genuinely autonomous evolutionary process.

Engineering Design Methods

Engineering design methods[8] underpin one area of cultural change in which there is apparently no question that developments depend on human creativity and purposeful intelligence. The implication is that technological "evolution" is a mere metaphor; but is this really the case? This section explores the methods that are involved in engineering design – a process which the engineer Ken Wallace characterizes as "converting an idea or market need into the detailed information from which a product, process or system can be made".[9]

Wallace emphasizes the need for a systematic approach to design. Intuition, inventiveness and insight all play their part in what is, after all, a very human activity – but they are supported and enhanced by a disciplined methodology. Once the initial demand for a product has been perceived, the question arises how to meet it. As a general strategy for problem solving, it is useful to reduce complexity by splitting the overall challenge into manageable subproblems, to be tackled independently (though in context – solutions to individual problems will influence each other), and then combined. This approach is evident in each of the four stages into which Wallace breaks down the design process.

The first stage begins with market research, to discover a gap in a range of products. The example that I shall use throughout this section, in order to root Wallace's rather abstract exposition in reality, concerns a gap in the Brazilian vehicle market. Consider a situation in which several small businessmen and farmers in Brazil need a small goods vehicle in order to transport their products and tools – but no suitable vehicle is currently available.

Importantly, there will be no *one* "correct" plug for this hole in the market: design problems are by their nature open-ended, although some solutions will of course be better than others. The best way to begin to achieve an acceptable solution is to define the task in a clear "problem statement". What Wallace calls "divergent thinking" (i.e., an open mind) will be used in preparing this statement: information is gathered from a variety of sources, and considerations raised from disciplines other than the one particularly relevant branch of engineering. A solution-neutral statement of the problem can then be formulated, in order to identify the true needs without making assumptions about how they should be met. In the case of the present example, a first version might be something like: "to provide transport for small amounts of goods over poor quality roads". Here it is not possible to be much more solution-neutral than this: such a problem would only be addressed by a car manufacturer, and therefore certain criteria are unavoidable (the new vehicle will use roads, will not fly, etc.).

After this, "convergent" thinking will be used to elaborate the target specification: the designer needs to limit the search field by detailing the precise requirements and constraints. Relevant considerations will be function, safety, economics and time scales. With the resources available, compromises often have to be made in later stages of the design, and one way of focusing on the best compromise is to have identified requirements, at an earlier stage, as "demands" or "wishes". Demands provide criteria for selection: they *must* be fulfilled, or the solution scrapped (e.g., meeting the relevant government regulations). Wishes provide criteria for evaluation: they are desirable but not essential (e.g., exceeding the regulations).

Having clarified the task, the second stage is conceptual design: generating concepts with the potential to meet the requirements. Solution principles will be created for all the subfunctions of the product (e.g., vehicle type, engine), and studied to see which can be combined with each other. Ideas will be generated via brainstorming, the study of existing devices, and in addition "useful ideas can be obtained from the study of natural systems."[10] Here, once again, divergent thinking will be used – this time to generate as many ideas as possible. Going back to the example, there is already so much established (no car manufacturer starts from scratch) that there are limited options for innovation. Typical choices will be between a pick-up or van, and whether or not it should be based on an existing vehicle.

Following this, convergent thinking will come back into play, as the best solution is selected. The "pass" criteria detailed in the target specification will be used to evaluate the possible solutions. Combinations of subfunctions will be scrapped if they fail to meet a demand, and the remainder evaluated against the wishes (weighted according to importance), with a view to determining which will "provide the maximum competitive advantage".[11] Notice that here, as at *any* stage of the design process, cost analysis may override engineering considerations.

Following the first two stages of the process, the selected solution must be presented to other people in a way that convinces them to move it to the next stages. This third stage is known as embodiment design, in which the concepts undergo a structured development. In the case of vehicle design, layout drawings and clay models, etcetera, will help to reveal which concepts won't work in practice. Again there will be a trade-off between the divergent thinking needed to suggest possible ways of meeting the target specifications (e.g., choices between different engine sizes or suspension layouts, front/rear wheel drive, etc.), and the convergent thinking needed to select between the possibilities that do meet the demands, on the basis of the wishes that each meets.

The final stage is detail design: specification of the shape, dimensions, materials and tolerances of each component. Again these will be evaluated against the target specifications – in a vehicle's case, by testing prototypes and using computer aided engineering (CAE). Thus it can be seen that at this stage, as in all the others, the design process is iterative. There are feedback loops between evaluation and details, perhaps even going back to the embodiment stage, if deeper problems arise.

In summary, then, the design process – which seems prima facie to be the harnessing of imagination to practicality – is underpinned by a methodology that is iterative and in many senses even mindless. It moves from a perceived demand, through clarification of the problem in a solution-neutral statement, and the generation and initial selection of concepts with the potential to meet the requirements, to a structured development and detail design of the end product. At each stage of the process selections will be made between possible solutions, according to the demands and wishes laid down in the target specifications. An option may be rejected when it is still an idea ("How about a pick-up truck?"), whilst it is being developed as part of the embodiment design ("Perhaps a 1.3 litre engine will give us the power we need"), or even when it has reached the final stage of detail design ("Let's try the engine from our existing pick-up in the prototype"). It may be rejected on the basis of economic as well as engineering considerations. If at *any* point it seems that the end result will not be viable, then losses will be cut and the project abandoned.

"Design Evolution"

It is clear that there are analogies between this design process and biological evolution. For instance, depending on the "pass" criteria laid down in the problem statement, a design evolves via an iterative process of divergent and convergent thinking in the next three stages – and this is analogous to the way in which, depending on the fitness criteria that are laid down by the environment, a species evolves via natural selection. Such analogies are interesting to explore, and give the sort of intellectual satisfaction that is the result of discovering any familiar pattern in an apparently different field. To what extent, though, is it justifiable to pursue the "design evolution" hypothesis? Is it helpful to talk of engineering designs "evolving", as though that process really did mimic the evolution in the natural world? In order to discover whether Darwinism illustrates a process that is also displayed in engineering, I need to know whether I can meaningfully apply to this area the key elements that have been worked out for the meme hypothesis in other cultural fields.

Possible examples of design memes might be the concept of a cantilever, the idea of using concrete as a road surface, the design of the "whale-tail" on a Porsche 911, or a particular way of using a CAE package. The significant fact about any of these examples is that the meme is the *information* contained in the blueprint for a design, rather than the end product itself. Just as in biology we refer to genes "for" bodily features (e.g. blue vs. brown eyes), so in engineering we might speak of design memes "for" artefacts' features (e.g., torsion bar vs. leaf spring suspension). The 911's whale-tail, for instance, is the end product of a successful design meme "for" a whale-tail.

Notice that, although design evolution may for convenience be referred to as "analogous to" genetic evolution, as in any area of memetics this should not be taken to imply that the former is theoretically dependent on the latter. Rather, both are examples of a more abstract, generally applicable theory of the evolution of replicators under conditions of competition. The two processes have the same description at a sufficiently functional, abstract level. Nonetheless, because we are already familiar with genetics, we can use it to illuminate what we might call "design memetics". In other words, although we should not expect the particular details of biological evolution to carry over into design evolution, it seems reasonable to exploit our knowledge of neo-Darwinism as a guide to what the essential elements of design evolution might be. Design memes have in common with genes the fact that both embody information which is replicated, varied and selected, producing a form of evolution that is observable in their phenotypic effects.

Yet of course this does not answer the question whether there are any grounds for *accepting* the hypothesis of design evolution. Are there aspects of engineering design that can realistically be characterized as replication, variation and selection?

It is uncontentious to suggest that there is variation amongst the designs that engineers produce. Even within the same model of a car, for example, there will be several choices to be made by the potential customer: type of fuel, size of engine, colour, extras and so on. Moreover, as part of the design process, engineers use divergent thinking in order to produce as many options as possible in their search for a solution. Thus variation is apparent not only in the end products, but also in the concepts that arise in the design process. In fact, the open-ended nature of design problems ensures that there will always be variety amongst the solutions proposed.

Recall that in nature variation occurs through the mutation and recombination of genes. Recombinations are limited by the need for alleles to correspond, resulting in a range of possible recombinations that is limited, though rich, with respect to any given gene pool. Genetic mutation is random with respect to increased fitness, although the mutations that can occur are limited by the nature of what already exists – for example, by genes' mutation rates and by embryology.

Clearly, mutations in engineering concepts are also "random" – not in the sense that they spring, as if by magic, into the engineer's mind, but in that they are random with respect to their "fitness" for the target specification. If this weren't the case, then an engineer would be able to latch on to the appropriate solutions straight away, without the time-consuming and costly business of testing them at the embodiment and detail design stages. Moreover, just as the consequences of genes' mutations are limited by the relevant embryology, so the effects of design mutations will be limited by the processes of translation into reality. An engineer employed by a major car manufacturer will be restricted in the innovations that he can incorporate in his designs by the existing manufacturing practices of that company. For instance, in the case of most major car manufacturers he can design fibreglass vehicles until he is blue in the face, but he will not be able to have them built. Clearly, too, recombinations of existing engineering concepts may be responsible for a new overall design – and just as genes must correspond with the alleles that they replace, so an existing concept may only be replaced by one that controls the same aspect of reality (trivially: the engineer may replace his vehicle's small engine with a larger engine, but not with a spare wheel).

Given that there is variation amongst these purported design memes, are there methods of transmitting them which might be regarded as replication? There are two aspects of replication: the preservation and the transmission of information. Engineering designs are preserved in the forms of blueprints, of prototypes, of CAE models and even of ideas in the minds of individual engineers. Such representations of information must fulfil various conditions if they are to count as replicators: they must, for instance, be able to interact with other such representations, and to exert some form of control over their environment. They are transmitted by being taught, mimicked, communicated, learnt... all of the usual processes of cultural transmission. There is nothing controversial here. Designs are replicated in the minds of the general public via advertising, and in somewhat more detail in the minds of other engineers via blueprints, and so on.

Perhaps more controversially, designs need to be particulate if they are to count as replicators and be subject to evolutionary change. If replicators blended with each other, then evolution by selection would be impossible. Once more paralleling memes in any other area, the hypothesized design memes can largely be counted as discrete on the same grounds that Mendel decided that the factors controlling his pea-plants' characteristics were independent and indivisible: observation of their effects, which are either present or absent. Each time it is replicated, an aspect of design runs the risk of embellishment, corruption or diminishment, but this is no different from the risk that each gene runs of mutation during replication. With reference to cultural evolution, engineering seems to be an area in which it is especially easy to observe the definite presence or absence, in an artefact, of any given design. A car either has or has not air bags, drum brakes, front wheel drive, for instance.

Evolution needs not only replication and variation, but the replication *of* variations, to offspring. Clearly, "offspring" does not here refer to biological but to cultural descendants – and equally clearly the variations are so transmitted. A young designer will be influenced in his practices, and restricted in his starting points, by the company that he joins. Furthermore, just as the variations that you inherit from your biological parents may develop differently in you, depending on the nature of your environment, so the variations that you acquire from your cultural predecessors (such as more experienced engineers, or lecturers) may develop differently in the context of your mind and environment. A safety engineer who hears about a novel innovation may decide, after analysis, that it does not improve the vehicle's crashworthiness – but still he is aware of its existence and able to retransmit that information. (If he tells me that this particular idea doesn't work, then he also tells me that it exists.) What matters, from the point of view of evolution, is simply that variations are replicated.

In design as in biology, then, variations exist and are passed on to the next "generation". In order for evolution to occur, the third factor that design needs is selection. What is it that a design meme needs in order to be successful? It will be popular and long-lived if it meets the various criteria laid down by the humans who want to make use of it. In other words, like any other meme it ultimately depends upon human beings' attention. Without this, no effort will be put into moving it from the conceptual to the embodiment and detail design stages – or even if it does make it so far, the consumers at whom it is aimed will not select it from the many alternatives at their disposal.

Design memes' competition for attention seems to be a consequence of the open-ended nature of design problems. A design will be long-lasting and widespread only if it succeeds in capturing the attention of enough people, to the extent that they regard it not only as a worthwhile focus for their money, time and effort, but also as a *more* worthwhile focus of attention than its rivals. Factors that come into play in their decision may include its compatibility with existing features of their lives (from garage size to self-image); the relative importance of those existing artefacts, opinions or practices; and the external environment. Artefacts' capacity to survive and be replicated is affected by their efficiency (or at least their *perceived* efficiency) in fulfilling their intended use.

The "fitness" constraints that are imposed on any particular design will be laid down by the initial problem statement, which specifies the conditions that a design must meet if it is to succeed in the practical and commercial worlds. Convergent thinking will then play the part of selection, as the engineers choose between their possible solutions. Demands may be compared with "life or death" criteria in the biological world, and wishes with the conditions that will determine an organism's quality of life: it will not live without meeting the demands; it will do better or worse than its rivals as a result of the "wishes" that it fulfils. Moreover, just as the engineer's ideas may never see fruition if they are overruled by considerations from other disciplines (e.g., economics), so a genetic mutation may fail to be translated to the phenotype as a result of embryological restrictions. In particular, recall the emphasis that is placed again and again by writers on biological evolution, on the fact that natural selection is never forward-planning: if a mutation is harmful *now*, then that organism may die before procreating, and the mutation will never be selected (even if in the long run it may have been helpful). Similarly, if at *any* stage a design doesn't meet its budget requirements, then everything stops (even if in the long run it would have been the best engineered design).

So it seems that there is competition between design memes for the limited resource of human attention. Added to their variation and replication, this will ensure that a form of evolution is played out in engineering design. At a far greater pace than genes, design memes vary, are replicated and selected – and thus they evolve. The preservation of those designs with the best fit to their environment, and the extinction of those without, should be expected.

Evolution and Design Reconciled

Human design methods, then, are evolutionary. Ideas and designs are reproduced, vary and are selected according to the relevant criteria, and the result is a panoramic variety of increasingly complex human artefacts. The culmination of this process is now being attempted in research establishments around the world: can artefacts be designed to display intelligence or even consciousness?

Thus the preceding account of the design of human artefacts provides a working, observable example of the compatibility of evolution with design – not just in principle, but in practice. Human design can be described in one of two ways. In intentional, psychological terms, the new front end for the model "xyz" car was designed by Chris because he wanted to make the "xyz" more crashworthy in frontal impact, he wanted to keep his job, and so on. This provides an answer to the "why?" questions that might be asked about the novel design. The "how?" questions, though, are answered rather differently. The new front end was designed using the four-stage process described, through which novel designs for that part of the car were tested against the "pass" criteria laid down in the problem statement. One of them was selected from the variety of proposed solutions, and the end result is a front end that is intellectually descended from, though a significant evolutionary improvement upon, the existing design.

For the claim that much of this process is mindless, it does not matter that the evolution of design memes is dependent on human minds. This is simply because thoughts and other representative media (e.g., language, blueprints) are the province of memes. Without the active stimulation of human minds, design memes may find safe havens in these media (in libraries, perhaps), but will neither replicate nor evolve. This is no different from the fact that genes are the units of biological selection, but depend on interaction with the environment and the mechanisms of embryology, in order to replicate and evolve.

It would appear, therefore, that there is no contradiction at all between the following two statements:

(a) The front end has evolved so as to fulfil the "pass" criteria laid down in the problem statement. That evolution may be described as mechanical: given the problem statement, an iterative process of selection determined which design will be chosen from the suggested options.
(b) Chris's purposes in creating his design are not reducible to any of the following: the problem statement; a description (no matter how detailed) of the means by which the front end was either designed or built; a description (no matter how detailed) of the front end itself. We have to seek out Chris himself, if we want to discover his purposes.

Thus it seems clear that evidence for memetic evolution is not the same thing as evidence against the human mind.

Different Points of View

Yet it should be noted that Dennett and Blackmore are not alone in their stance on this matter. Others have agreed that we must "reject the notion that some 'central executive self' can pick and choose among the memes, and refer instead to the sorts of filters (cf. Dennett, 1995) which memes and genes have constructed".[12] The author of this quotation, Nick Rose, adds even more bluntly that "if variation among memes is somehow directed by consciousness towards some goal then it is not a Darwinian process."[13]

The latter comment was a response to the suggestion that the artificial selection of domestic animals could be seen as a process inexplicable "without reference to selves, goals and intentions".[14] I would argue that it is, rather, a process which can be used to illuminate the crucial issue of the different "points of view" of the elements involved in evolutionary processes. In artificial selection, the "selves, goals and intentions" of the humans involved are – from the genes' point of view – simply one aspect of the environmental pressures on them. From *our* point of view this process is inexplicable without reference to our goals, hopes and fears; but from the *genes'* point of view this situation is no different from any other. They replicate, they vary and some fit better to the environment than others. The fact that humans are shaping that environment does not undermine the essentially Darwinian process that is unfolding.

Similarly, from the memes' point of view the conscious direction of human minds might be seen simply as a part of the machinery of the cultural evolutionary process.

The distinction between this outlook and the Dennett-Blackmore hypothesis is that I acknowledge the validity of *three* different points of view: genes', memes' and *ours*. From the genes' point of view, they are struggling for survival in an environment that consists in a variety of elements: other genes; external factors such as the physical world and other genetically built "survival machines"; the environmental changes that have been effected by humans and their memes. From the memes' point of view the environment consists in other memes; external factors such as the physical world and the existing cultural environment; genes; and our minds. From our point of view the environment consists in memes; genes; other people; our physical and cultural surroundings – and the existence of our point of view need not affect the Darwinian nature of what is going on from the perspective of either genes or memes.

On this theory, Dawkins is right to claim that "we have the power to defy the selfish genes of our birth and, if necessary, the selfish memes of our indoctrination.... We, alone on earth, can rebel against the tyranny of the selfish replicators."[15] Genetics is concerned with the development of nature, and memetics with the development of culture. The development of human individuals is a separate topic for both realms, even though our development is of course dependent on the existence and support of both underlying evolutionary mechanisms. Just as atoms underpin all that is physical, but from the perspective of the interactions between medium-sized physical objects are not what matters, so genes underpin all that is biological but from the perspective of the interactions between their survival machines are not what matters, and memes underpin all that is cultural but from the perspective of the interactions between minds are not what matters. In each case there is a unifying foundation, but also another, emergent level of description that is not only equally valid but in some contexts by far the more useful.

Conclusion

Memes, then, are separate from the mind, which is neither composed of a meme complex nor built by memes in the way that genes build their survival machines. Rather, the mind has a certain innate potential which develops as a result of interaction with its environment, both physical and cultural. Some of what it takes in will indeed be absorbed so deeply that it could accurately be described as a part of that individual, forming a filter for future cultural input (counsellors sometimes refer to this as a "script"). Nonetheless, we retain a degree of choice and flexibility in our reactions to many of the experiences and memes that we encounter, as is shown when siblings form surprisingly incompatible memories of incidents and themes from their shared childhood.

The consciousness and intentionality that is being claimed for our mental lives is, however, quite reconcilable with the essentially mindless process played out in any evolutionary algorithm, whether its medium be biological or cultural. In many areas of science different levels of description can be applied to the same phenomenon, and it is no different here. The world can be seen through physical, chemical, biological, cultural or psychological lenses, and the mindlessness of the cultural evolutionary algorithm need no more undermine our identity as conscious selves than does the mindlessness of the physical or chemical descriptions of our interactions.

13

Science, Religion and Society: What Can Memes Tell Us?

Having tested the structural foundations of memetics, there may now be further benefits to be gained by looking at some of its more practical applications. This chapter aims to deepen our understanding of memetics, as well as of science and religion, by examining those cultural areas through its lens.

Science

There are many cultural areas in which knowledge and skills are passed on and develop between "generations", but perhaps the most notable is science. How does it look, from the perspective of meme theory?

The most obvious starting point is the emergence of novel theories. Innovation, according to meme theory, is due to two factors: recombination and mutation. In recombination, existing memes are appropriately recombined in new situations, creating new ways of thought and novel effects, perhaps as the result of previously recessive memes' "effects" being revealed in the reshuffle. This sort of memetic innovation is seen, in science, in the process of extrapolation from existing results to a novel theory. Existing theses are reshuffled – perhaps in the light of new evidence – and this may lead to unforeseen consequences, or even to a fresh hypothesis.

This process cannot, however, account for the "eureka!" phenomenon, where the hypothesis was not itself the direct outcome of previous results. Such instances appear to be more in keeping with the mutational element in memetic variation. If this is so, and the mutation of memes is a good model for scientific innovation, then what does it predict about the nature of such innovation? First, it must lead us to expect that new theories are as likely to be false as true, and as likely to be incompatible with existing evidence and thought as not, since the most important aspect of memetic mutation is its randomness: the fact that it has no intrinsic bias towards increased fitness. Clearly, this expectation has been met. Just as many genetic variations actually *decrease* fitness, so a large proportion of "eureka!"–type shrieks are followed more or less closely by expletives: the new idea does not have the explanatory or predictive success necessary for the survival of scientific hypotheses, or it is rejected as incompatible with contemporary beliefs.

It has been noted, however, that there are restrictions on mutations' randomness: it is not true that any convenient mutation might occur. Which memes do mutate, and in what way, will be constrained by their content, by their environment and by the existing "embryology". To put this in terms of scientific theories, their development will be constrained by their subject matter, by the best available evidence and by the consequences of existing thought in that area. None of these observations is particularly original, but they should lead us to look at the development of science in a different way. There is a popular view that the progress (even if slowly and not very directly) of scientific thought towards the truth will be constrained only by the limits that technology imposes on the best available evidence. In contrast, meme theory implies that some of the most significant restrictions on scientific progress will stem more from the *existence* of whichever theories, evidence or methods there are already, than from their accuracy and suitability.

Indeed this provides a neat account of the historical success of theories which we now view as farcical (consider phlogiston's "negative mass", for example). It does not seem likely that our intellectual predecessors were less intelligent than us; the deficiencies that we now see in their theses must, therefore, also have been available to them. According to memetics the reason why these problems come to light now, although they did not do so earlier, is that the meme which prevailed at any particular time was the available meme that was then most compatible with the rest of the meme pool. Indeed, it would not have *been* so successful if that were not the case. Now, on the other hand, many of the past's memes are not compatible with the existing meme pool, and we see problems with them as a result of their conflict with prevailing ideas (including the latest available empirical evidence).

In this context it is worth remembering that, like genetics, memetics is simply a theory about the transmission and development of information.

It accounts for the relation between that information and its external consequences, but says nothing about its intrinsic value. Biological fitness is a relative concept: allele *a* may be selected rather than *b*, but this does not tell us whether *a* is really an efficient method of survival and propagation. For the answer to that question we have to appeal to theories other than neo-Darwinism (engineering, for instance). Similarly, if idea *x* is selected rather than *y*, then this tells us nothing about the truth, elegance or other values of *x*. For that information, we have to appeal to other theories (aesthetics, for instance, or truth criteria). In science as in other areas of culture, then, we should expect the successful (i.e., long-lived and fecund) ideas to be the ones that are more fit than their contemporaries for the current cultural environment (i.e., better tested, more compatible, etc.); we should not expect this relative success automatically to be a reflection of their accuracy.

On the other hand, the existing cultural environment will tend to increase in "volume" over time: unless no historical records are kept, knowledge acquisition is a cumulative process. So although it is true that if the existing theories are wrong then it will be hard to escape their legacy, it is equally true that if they are (even approximately) right then they should act as a springboard to deeper knowledge. Dependence on the existing meme pool – and, as for any type of evolution, on chance and error to provide at least some of the mutations – does not, therefore, have to be grounds for total pessimism about the possibility of progress.

When scientific progress does occur, many would agree that it is truly evolutionary – in the sense that it takes the form of a pattern emerging from conflicts between individual theories. This point is made by Shrader,[1] who also comments that such progress is unlikely ever to be a process of strictly rational consideration of the evidence and alternatives. There is so much published each year, in every academic discipline, that no scientist can hope to read even just the well-researched articles. As emphasized throughout this book, then, a novel theory will have to find some way of grabbing the relevant scientists' attention: if it is perceived as irrelevant or too far-out, then it will simply be ignored. This also highlights Shrader's point that most (accepted) discoveries are made in the context of a continuing tradition: if they really were as completely radical as they are sometimes portrayed in the history and philosophy of science literature, then they would probably be dismissed out of hand.

The structure of the scientific community will also affect the selection pressures on theories: the professional standing of an individual scientist will have a bearing on the reception of his work, and politics will affect funding and thereby the progress that can be made in any given discipline. The lifespan of novel scientific theories may well be affected by such factors. Once again, though, this is not grounds for total pessimism. Shrader also notes that no amount of funding or professional kudos can make a false theory true.

Thus the memetic perspective on science reveals little that is really surprising, but it does help to demystify some of the processes at work. Somewhere between the traditional view of scientists as invincible warriors in the battle against ignorance and confusion, and the more recent cynicism about their relationship with government and other vested interests, comes the claim that their work is but one branch of cultural evolution.

Meme theory highlights, in particular, the interaction between the subjective world of scientists and the objective world of hypothesis and evidence. At one level, science can be seen as a system in which novel ideas emerge via the recombination and mutation of existing hypotheses and are subject to selective forces such as the very existence of those current ways of thinking, as well as politics, funding availability and ad hominem considerations. Crucially, however, one of the strongest selective pressures on scientific theories is their compatibility with the evidence. The fact that the best available theory will always be just that – the best that is currently available – need not undermine the scientific enterprise, whose mission is to match theory to reality, and whose methods have been honed to ensure that, on the whole, it succeeds.

At another level, my version of memetics allows science simultaneously to be described as a system run by human scientists. Their rivalries, egos and respective statuses may well be the source of selection pressures on the hypotheses that they create, but at the human level this is not what matters. The human beings engaged in science have a whole range of priorities, such as intellectual satisfaction, emotional fulfilment and peer respect. They are people like any others, with career aspirations, home lives and principles.

They also have the human ability to meta-represent. They can step back from the scientific enterprise, taking its theories and their implications into completely separate representational arenas like philosophy, history, theology and morality: What is the aim of science, and how does it progress? How have its theories developed over the centuries? What, if anything, can it tell us about religious matters? How ought its results to be implemented, and what safeguards need to be put in place? Questions like these matter, and the fact that *science* can be described by a theory of cultural evolution does not absolve *scientists* from their human responsibilities. The view of memetics that I have defended in this book allows both levels of description their place. As science evolves, towards what we hope will be the most accurate possible *representations* of the world, so there must be a parallel process of *meta-representation*, whereby we reflect on its methods and outcomes.

Religion

Another cultural area that has seemed to many memeticists an obvious target for the application of their theory is religion. In particular, those like Dawkins and Blackmore who are themselves atheists have seen the meme hypothesis as another arrow in their quiver of ammunition against God. As noted in Chapter 8, in Dawkins's view religion is more like a mental virus than a "good" meme, and his theory of cultural evolution can thus be extended to explain away religious belief as "an infectious disease of the mind".[2] Extensive arguments were ranged against Dawkins's virus-meme distinction and there is no need to revisit them, but his specific overapplication of memetics is far from the only way in which it has been claimed that religion is undermined by any close examination of human culture.

For instance, one of the key factors that leads many people towards scepticism about religion is the close correlation between an individual's background and his religious beliefs – and indeed memetics emphasizes the significance of the cultural environment for the success of particular memes. It seems obvious that a Christian culture breeds Christians, and conversely that Christian people are likely to have a Christian background. As Bertrand Russell put it: "With very few exceptions, the religion which a man accepts is that of the community in which he lives, which makes it obvious that the influence of the environment is what has led him to accept the religion in question."[3] The claim is that people do not, on the whole, take up a religion because they have been moved by rational argument: "most people believe in God because they have been taught from early infancy to do it."[4]

Of course it would be ridiculous to say that nobody ever becomes a Christian who is not raised in a Christian society: people have converted to Christianity from all manner of faith (and lack of faith) backgrounds. Equally, there are plenty of people who spend their formative years within a Christian society, and yet adopt a different belief system as they mature – whether Islam, atheism, or whatever. Indeed, clergy children are notorious for rebelling and taking a different path from the one that their parents have chosen. The chances are that a person who was forced to attend Mass each week as a teenager won't darken church doors again for years afterwards.

Yet the opposite can be true, too: sustained cultural opposition (under a Communist regime, for instance) cannot stifle people's faith. Consider, also, the fact that social reformers will by definition be going against prevailing practice through the influence of conscience – which cannot therefore be just the result of prevailing practice.

Clearly, then, this argument is not as straightforward as some of its proponents imply; indeed their very existence demonstrates as much. The claim that Anna is a Christian because she was brought up in England, by parents of Christian extraction, sounds strange when it is made by an atheist who was brought up in England, by parents of Christian extraction. Why should her background be an adequate explanation of her belief system, but the same not be true of a nonbeliever? Or perhaps the same *is* true of the nonbeliever, whose atheism is simply a product of a more sceptical infant environment? "A student once criticized Dr Frederick Temple, then Archbishop of Canterbury, saying, 'you believe what you believe because of the way you were brought up.' Temple replied, 'That is as it may be. But the fact remains that you believe that I believe what I believe because of the way I was brought up, because of the way you were brought up'!"[5]

It is notoriously difficult to pinpoint the reasons behind people's opinions – especially those as emotionally significant as their religious beliefs. The crux of this argument, though, does not depend on an endless "'yes you are,' 'no I'm not'" response to the claim that most people are only Christians because of their background. Rather, it turns on the assertion that our reactions to religious matters are peculiarly different from our reactions to other controversies: whereas we usually check our beliefs against criteria of rationality and evidence, our religion is so much a part of our cultural heritage that we don't bother with these checks – it just seems right to us. Religion is, in addition, so emotionally significant that most of us will resist any attempt to challenge it. The argument with which this section is concerned, therefore, is that religion is not only the product of our environment – after all, lots of things are – but that the environment moulds our religious beliefs in a disturbingly pernicious fashion.

Memes, it has been noted, are subject to selection pressures which vary according to their content. A meme for a scientific theory, for example, might be favoured because it corresponds well to the evidence, facilitates useful predictions and does not contradict the rest of scientific knowledge; a meme for a melody might do well because it excites pathos; a meme for a particular design of bridge might succeed because that type of structure is robust as well as attractive.

When it comes to religion, the claim is that its successes are not based on rationality. Rather than being compatible with other existing theories, or standing up well to independent verification, religious ideas do well simply because they have found some way of ensuring that they are taught and imitated. The popular and long-lasting religions are the ones that are emotionally appealing, find ways around our rational defences, and include the claim that it is necessary to pass them on to others. Thus the selective pressures in spiritual matters bear no resemblance to the criteria of rationality and evidence to which scientific theses, for instance, are subject. In science, we decide which theory to adopt on the basis of superior evidence, but in religion we just follow our parents, or possibly some "particularly potent infective agent",[6] as Dawkins puts it, whom we happen to encounter. Informed decision between the world's faiths plays no part in this process of selection. Religious memes are unique in our culture, in that they are able to bypass some of the filters through which we usually process new claims.

One of Dawkins's main arguments for the fact that religion is a mental virus (or at least unique amongst memes), is that those who have been infected by it, but have now "recovered", still insist on infecting their children with it. The best explanation of this, he says, is that the virus has thereby provided itself with future victims, even if earlier ones recover. Dawkins derides the reasons given by such parents for having their children baptized: he says the claim that children deserve the *choice* of whether or not to believe is, at best, a good argument for telling them about *every* world religion.

Even disregarding the weakness of his meme-virus distinction, however, the reasoning here is faulty. Religious parents rear their children within a faith because they regard *its* teachings as *true*. Atheistic parents, on the other hand, regard the teachings of *all* religions as *false*. Therefore if they decide to provide their children with the background necessary to make an informed choice about whether to adopt a religion, then they are unlikely to care, particularly, on which religion their offspring are deciding. It is only through open-minded generosity that they are willing to give their children the option of believing in *any* such fiction. It would be clearly impracticable for them to educate the children in *every* faith – and to do so would be to spend a disproportionate amount of time on religion, considering that the parents under discussion are atheists. Thus the only ground for their decision between the world's religions will be their own familiarity with the one in which they were raised. Indeed, this familiarity may be the very reason for passing it on: Christianity, for instance, is part of the historical fabric of British society, and it may be that nonbelieving parents simply want to give their children an understanding of this aspect of national life.

Moreover, such parents' actions may more simply be explained as a result of open-mindedness than as the result of a blind obedience to instructions that were part and parcel of a long-discarded belief system. As noted several times already, we possess lots of information to which we do not subscribe, and which therefore exerts little executive control over our thoughts or behaviour. Religion need not be seen as an exception to the general rule that no open-minded person objects to his children's adherence to beliefs and tastes that he no longer shares (unless he has particular reason to regard them as harmful). Thus he might pass on to his children books that he has bought but did not enjoy, or records of bands whose music he no longer appreciates – and usually these will, inevitably, reflect his own culture and background. He might tell them about the political opinions that are opposed to his own (usually this would, at least initially, be restricted to those of their own country), or explain why some people hold different beliefs about various moral issues. There is no reason why religion should constitute a special case: for an atheist, it is merely another set of beliefs to which he does not adhere, but about which he wishes to say to his children something like "there are people who believe this, for these reasons; I don't; you may decide for yourselves."

Nevertheless, in support of the claim that religion is adopted for emotional and not rational reasons, the tendency of some religious people towards fanaticism and gullibility, when it comes to their spiritual beliefs, is often highlighted. Surely such characteristics strongly imply that religious memes have some sort of unique bypass around our usual systems of reason and logic? Yet neither trait is in fact unique to religion: both can arise in any area in which the beliefs at stake are important and/or life changing. The scientist's "eureka!"–type experience is itself not wholly devoid of emotion. Does this mean that what we perceive as exciting scientific discoveries may, rather, be cases of infection by a mental virus that exploits the "internal sensations of the patient"?[7] That the keener the scientist is on his new theory, the more evidence there is that it is really a virus? Obviously not, any more than for people's emotions about their religion. There is no reason why one's feelings about a claim should imply anything about its truth value. This is the case for the ways in which one arrives at beliefs about all sorts of information: in science, "eureka!" is as valid as years of hard slog if the end result is correct, and in religion hearing it from your parents is as valid as working it out for yourself or by revelation, if what you get is the truth.

This last point hints at a broader distinction between the world and our attitude towards it. On the one hand there is some objective truth about the nature of the universe, our place in it, and whether or not God exists. There is a "fact of the matter", if you like. On the other hand there are questions about how (or indeed if) we can discover the facts of the matter, and how we feel about what we learn. For instance, Dawkins finds it surprising that we are especially likely to share our own parents' religion, or that of the culture in which we are raised: "since religious beliefs purport to be true all over the universe it is odd, to say the least, that which belief you hold depends so heavily on where in the world you were brought up." But why is it odd? Rather, it is perfectly reasonable that people's views of the universe should be based on their place within it (i.e., be culturally founded). This cultural dependency of our *beliefs* is quite separate from the objective truth of the *facts*. The facts according to an atheist are quite different from the ones that a Christian or Sikh believes to be the case – but the truth value of each point of view is not affected by its cultural grounding.

That this is the case is actually quite fortunate for Dawkins, for should his argument be valid then it would also count against atheism. Many people absorb atheistic beliefs from their parents, and it is increasingly common for people in the West to be brought up within a secular culture in which religion is subject to ridicule or simply ignored altogether. Does this mean (as Dawkins implies that it does for other religious beliefs) that atheism must be false, since it is in these cases merely absorbed rather than chosen? Of course it doesn't: the ways in which we acquire a belief are completely irrelevant to whether it is true or false. As the well-known atheist George Smith puts it: "The American child who grows up to be a Baptist simply because his parents were Baptist and he never thought critically about those beliefs is not necessarily any more irrational than the Soviet child who grows up to be an atheist simply because his parents were atheist and because the state tells him to be an atheist."[8]

Again we can see that there is a distinction to be made between the evolution of claims and practices in this particular cultural area and the beliefs and priorities of the humans who engage in it. On the one hand religious ideas, like scientific hypotheses, evolve towards what we hope will be the most accurate possible representations of the world and our place in it. (Theists believe that religion is aided in this process by the revelations of a loving creator God, so that in this area at least it may be that cultural evolution is not wholly autonomous.) On the other hand these ideas are accepted or rejected by human beings, and it may well be true that some people set aside their usual capacity for rationality and restraint when it comes to religious matters, just as others do when dealing with politics or family feuds. As in any other cultural area, my version of memetics allows both levels of description to hold.

As a final consideration, I should note that a theme running through almost all of this section's arguments against religion is something that C. S. Lewis has termed "Bulverism".[9] He points out that before explaining *why* someone's views are wrong you first have to show *that* they are wrong. "The modern method is to assume without discussion *that* he is wrong and then distract his attention from this (the only real issue) by busily explaining how he became so silly." This indeed seems to be the method of many of the arguments from culture against religion: to *assume* that religious beliefs are unfounded, and use cultural heritage, individual fanaticism or viral infection as an explanation of why people hold them regardless.

The fact is that, although our environment undoubtedly plays its part in shaping our faith, there is really no basis to the claim that culture's influence on our religious beliefs is such as to undermine their credibility. Memetics is a theory about the development of ideas and information: it has little if anything to say about the truth or falsehood of either religious beliefs or indeed any other sort of belief. We are able, as explained in the previous chapter, to make choices about what we do with the cultural baggage that previous generations have left us: to use our capacity for meta-representation in order to collate and evaluate it. The route via which religion was handed down to us is irrelevant to the question of its validity.

Contradictions from Genetics

Despite the irrelevance of memetics to the question of religious truth, however, we have seen that in its application to the development of the natural sciences meme theory appears to enjoy a degree of explanatory success. Are there any other areas in which it might usefully be applied?

A significant piece of evidence in favour of gene theory was its ability to explain various apparent empirical contradictions, such as altruism. Is meme theory able to explain apparent contradictions that arise from gene theory, such as suicide or contraception? In fact, this seems to be one of the theory's strongest candidates for success. Clearly, a *gene* for suicide, self-sacrifice or contraception could not replicate successfully without various complex strategies to compensate for its lack of fecundity; similarly, at the level of the individual such behaviour is inexplicable. Viewed in meme terms, however, such examples are easily explained.

Suicide

Consider an emergent meme for suicide: with the prior existence of a meme for heaven or at least the peace of death's oblivion, and circumstances which render life unbearable, the meme for an escape from life to death would surely be well received. Even if the recipient does not act immediately upon the information, the idea of killing oneself has surely been preserved in a form that retains its *potential* for a behavioural result. The deep shock that greets the news of a suicide would mean that such news would spread very quickly – so the meme for suicidal behaviour would not only be well received within the existing meme pool but also highly fecund. Hence, even though suicide destroys the individual bearer, it promotes the survival of its meme. For a deeply unhappy person who has this meme, and is looking for a way out of a situation when he has lost all hope that it could be changed or improved, it is easy to move from feeling that he cannot bear to live *like this* any more, to the belief that he cannot bear to *live* any more.

This is not to say that someone might kill himself simply because he had heard of suicide. Like any other meme, it could not exert its effects unless the circumstances favoured them. It is a firmly established principle within the Samaritans, for example, that asking someone whether he feels suicidal will not increase the risk of his killing himself. The suicide meme is in most cases recessive, in that it produces no effects, despite being successfully replicated. The point, though, is that this recessive meme can bear its tragic fruit when context renders it dominant. The most significant element of the "dominant" context is, I would suggest, the belief that there is no other way out of the person's situation. A suicidal person, asked whether she really wants to die, will often answer that what she really wants is an end to the situation or feelings that are making her suicidal – and that suicide seems the only way to achieve this. Helping her to come up with alternatives can change the context for her suicide meme, making it recessive once more.

There is an important corollary to this view of suicide as a meme that we can separate from its phenotypic effects. Contrary to some popular opinion, people who *talk about* feeling suicidal often *do* go on to make suicide attempts, and this is just what the memetic view would predict. If their previously recessive suicide meme has begun to be expressed in speech, then this is a hint that its phenotypic effects are beginning to be implemented – that there is currently a context in which it could dominate. Again, the opportunity to talk about why the person is feeling suicidal, and what the alternatives might be, can be enough to change that context back to one in which the suicide meme recedes.

Contraception

Another genetic anomaly is the ubiquity of contraception, and again this is a fact that benefits from a memetic explanation. The idea of contraception would be well received for various obvious reasons, and despite its adverse effects on the survival of its bearers' genes, its fecundity would be proportional to the number of people with whom its bearers had sex. Later, information about it would spread verbally, or via written professional advice, and at this stage its fecundity would be self-perpetuating.

Interestingly, although I have written that contraception will have "adverse effects on its bearers' genes", this is in fact rather too simplistic a view. As noted in Chapter 9, contraception will indeed have negative genetic effects, if there are sufficient resources available for any children who would have been born if it had not been used. In circumstances where resources are scarce, however, the meme for contraception could instead be genetically advantageous, so long as it is used to keep the population within the limits of its resources (although obviously this would not apply to any couples who used it to prevent their having any children at all).

The key in either situation, however, is that memetic success is independent of genetic advantage, and the introduction of meme theory can therefore account for some phenomena that genetics just cannot explain.

14

Conclusions

The theory of the selfish meme was introduced, some chapters ago, with a challenge: what does it contribute to our understanding of cultural change? I said then that my approach would be to focus on the underlying structure of Dawkins's hypothesis: to examine whether it could be true, is internally coherent and could form a solid basis for any empirical applications. What have those enquiries revealed?

The Meme Hypothesis

Ideas and customs develop at a pace that is far too great to be picked up at the level of biological evolution, and sociobiology's attempts to show how the evolution of the body could account for changes within our culture are therefore bound to fail. Richard Dawkins's suggestion is that we should look instead to evolution within culture itself, and he has proposed that this might occur via "memes", which are (roughly speaking) the cultural analogues of genes. On this view Darwinism is an example of a general type of theory which we should not artificially restrict to the realm of biology. Its essential features can be extracted and their domain of influence extended: whatever the type of replicator involved, their variation under conditions of restricted resources would lead to a form of evolution, and memes are simply cultural replicators. This is not to say that they will be tied to the particular pattern of development that genes have followed, for they are a *different* form of the type of process that Darwinism exemplifies – the term "analogy" should be used with great care in this context.

In order for Dawkins's hypothesis to hold water, the three key aspects of evolution (replication, variation and selection) must be shown to apply in the cultural realm. In particular, the existence must be demonstrated of identifiable units of replication that are realized in the appropriate way.

Replication: A Process of Assembly

Human culture is so vast that any theory of its development must be able to account for the ways in which its complexity has built up over the millennia. The most efficient methods of replicating complexity are hierarchical – or (to use a phrase less laden with distracting connotations) processes that build assemblies of subunits. If variation were permitted in every element of a complex structure then copying processes would lose much of their stability. As the constituents of our complex culture, memes must therefore be dependent for their replication on assembling constraints: this means, for example, that they must be able to slot into established cultural assemblies without their own informational content being lost or blended in the process; and whilst the results that they produce might be fixed, such packets of information must also have a degree of flexibility that enables those effects to be produced in a variety of cultural contexts. Copied in these ways, information is given the stability to grow and develop in complexity. The breadth and depth of human culture is thus explained by the cumulative replication of particulate information.

Particulate Memes

Yet the particulateness of human culture is one of the features of meme theory that has proved most controversial. For cultural transmission to benefit from the stability that assembling constraints provide, culture itself must be composed of particulate units of replication in the same way that the natural world benefits from the particulateness of genes, and it is not immediately obvious that this is the case. Yet there is the same evidence in its favour as Mendel once presented for his theory of genes: the clear presence or absence of the replicators' effects on the world. Just as genetic effects can sometimes give the misleading appearance of blending, so it can be hard in practice to separate the effects of one meme from another – especially since the actions that result from meme possession can be so very complex. This is not to say, however, that it is impossible in principle to identify units of cultural information – just that it is sometimes difficult in practice (as indeed it was for genes in the earliest years of genetic theory).

What does our understanding of cultural change gain from seeing it as a process based in interactions between particulate memes? At the most basic level the answer is that memes' particulate nature is what facilitates cultural evolution: the processes of variation and selection, on which any form of evolution depends, cannot take place without the heredity mechanism of discrete replicators. In addition, their particulateness is what supports the assembling mechanisms that provide culture with its stability and potential for growth: if bits of cultural information were constantly blended with each other then the distinctive features of each element would soon be lost. Even within the individual mind, the particulate nature of what we learn is what underpins our ability to partition existing knowledge, to organize and manipulate incoming information, and systematically to synthesize different parts of what we have learnt.

The important point to remember in memetics, just as in genetics, is that the replicators exist independently of their effects: even when their effects appear to blend, the replicators themselves remain particulate. They carry information *about* the effects that they control; they are not identical with those effects. In the case of genes, their independence is maintained via the medium of DNA, which preserves biological information in a form that is replicable and can produce its effects in a variety of contexts. In the case of memes, this role is performed by representational content.

Representational Content: Memetic DNA

What do we know about this cultural DNA? Many organisms are capable of representing the world around them, but only some of these representations will count as memes. Only organisms that are capable of a certain amount of behavioural flexibility, for example, will be able to form representations with a wholly determinate content: if it is not possible for the organism to adapt its behaviour in response to environmental changes, then it is equally impossible for us to determine exactly which bits of its environment are included in its representations. Without this determinacy of content, its representations cannot be memetic, because they do not perform the vital replicative function of *preserving* a given portion of content.

Other organisms are able to form determinate representations of the world, but not to learn from each other: in other words, the content of their representations is not replicable. Clearly these representations cannot count as memes – cultural *replicators* – either.

Memes, then, are representations which preserve their content in a way that can be copied between generations. As representations, they are specifically those bits of our mental "furniture" which control our behaviour in response to the information that they carry. In other words, memes' basis in representational content is precisely what accounts for their ability to exert executive effects on the world.

More than this, however, memes must be able to interact and assemble with other memes in order to account for the breadth and cumulative stability of human culture; and they must be able to represent highly complex portions of information in order to account for the depth and complexity of human culture. Again, these faculties are the result of memes' basis in representational content. Some organisms (namely humans) are able to form internal links between representations, in addition to their links with external perceptions and behaviour: these internal links give representations their internal properties such as identity, and ultimately free them from their dependence on external stimulation. Representations can now be meta-represented, and thus gain independence from their original context, as well as developing in complexity and abstractness.

As has been noted before, variation is necessary before a characteristic can be selected, and there is no question that modern humans vary in their abilities and tendencies to make links between the knowledge and information that they acquire. In particular, gifted children and adults typically display "keen powers of observation and reasoning, of seeing relationships, and of generalising from a few given facts".[1] In other words, these individuals are especially good at making links between ideas, of the sort that can ultimately result in works of creative or intellectual genius. Such observations lend support to the thesis that our ability to form internal mental links is both innate and key to the development of human culture – for not only does this faculty vary amongst humans, but still today it characterizes the brightest amongst us.

Just as genetics was limited at first by a lack of understanding of biological heredity, so memetics has until now been restricted by a lack of understanding of cultural heredity. And just as the nature of DNA provides the mechanism for biological heredity, so the nature of representational content provides the mechanism for cultural heredity. It explains how memes can preserve information between cultural generations in a form that enables them to exert their phenotypic effects in a variety of contexts, and thus accounts for the preservation and transmission of the information that constitutes human culture.

One DNA; Many RSs

The forms that representational content can take in its fulfilment of this role are various. Individual representations gain meaning from their context within a conventionalized representational system (RS) – this is as true for DNA as for cultural RSs – and the uniquely human capacity that lies at the heart of culture is our ability to copy and develop RSs as well as adding individual representations to our repertoire: the ability, in other words, to meta-represent. Natural languages, systems of mathematical and musical notation, the conventions of engineers' drawings – all are examples of cultural RSs, and each is peculiarly appropriate to its particular cultural area. Whereas organisms acquire their replicators in a job lot from their parents, human minds acquire replicators on an ongoing basis throughout their lives, and this means that they can acquire novel RSs as well as novel representations.

Amongst these various RSs, the natural languages have primacy: they alone benefit from an innate device for their acquisition. Yet they benefit, too, from the innate ability to meta-represent – and it is this which allows us also to develop nonlinguistic RSs, whose diverse rules and structures are realized in media other than speech. Once these sorts of RS have been taken into account, it becomes clear that there are many concepts that are not available to us until the RS that supports them has been developed. Our understanding of the complexity and enormity of human culture is thus increased: we can see that it is facilitated by the development of specialist RSs which enable the development of novel concepts, thus acting as a springboard to a growth in knowledge and the development of artistic expression.

Where Are Memes?

Representational systems are found both inside and outside human minds: we can manipulate representations in our thoughts, but are often assisted by writing things down as we go; vast stores of information are maintained in a variety of external media, although they depend on interaction with a human mind for their copying and development. More specifically, we struggle to use speech or thought alone as media for the realization of nonlinguistic representations, relying instead on the support of cultural artefacts like pen and paper. Whereas our language instinct gives rise to speech as the primary medium of natural languages, the universal grammar of those languages proves too restricting to allow that same medium to support nonlinguistic RSs. They are realized, instead, in external, physical media like books.

An important consequence of this range of cultural media is that the location of each representation will affect the ways in which it is able to preserve and replicate its content. Some representations can play an active role in ensuring the fecundity of the information that they carry, whilst others will have a more passive role, in ensuring its longevity. This picture of cultural change, as based on the hereditary mechanism of representational systems, can thus account for both its continual development and the remarkable persistence of its content.

The effects of cultural information, too, can be found both within and without humans: our behaviour is affected by information especially when it is presented to us in manipulative formats like advertisements; the effects of our thoughts and inventions are seen all around us in cultural artefacts like bridges and pianos. Yet if both memes and their effects can be found both internally, in human minds, and externally, in human culture, then what does this imply about the relationship between those minds and the culture that surrounds them? Is culture the product of human minds, or are our minds the product of the culture that they inhabit?

Human Minds and Culture

Most people's intuitive answer would probably be "a bit of both" – and the advantage of seeing culture as based in a variety of RSs is that this perspective enables us to affirm that intuitive response, and moreover to explain *why* it is valid. According to this view, humans are born with a degree of mindedness that includes, for example, the "representation instinct": an ability and tendency to learn and manipulate vast numbers of representations, as well as the various systems in which they are embedded. Humans are surrounded by such representations and their RSs from the moment of birth – the most pervasive being natural language – and infants' innate mental potential is realized as a result of exposure to this cultural environment. Conversely, the origins of those representations and their RSs are to be found in other, more fully developed human minds.

Yet culture does not "build" human minds in the same way that biology builds our bodies. Genes preserve and replicate biological information by building vehicles for their own propagation and protection. In other words, part of genes' job is to create the replicative apparatus on which they depend. Their effects may be found both within the survival machines that they build and on occasion in the external world (or even in other organisms) – but their replication depends ultimately on the machinery that they build for this purpose. Memes, on the other hand, depend for their replication on a faculty of the human mind that is at root genetic: the representation instinct. Although that faculty could not fully develop and play its vital role within cultural evolution without the acquisition of existing memes, this developmental process has more in common with the development of a muscle by use and exercise than with the ontogeny of an organism. Both organisms and minds develop as a result of the interaction between innate potential and the environment – and in the mind's case a crucial part of that environment is composed of memes.

A human mind, then, is partly the product of the memes that bombard it, but only because it has the innate potential to interact with and develop in response to those memes. Culture, in turn, is ultimately the product of human minds, but the preservation of information in representational content ensures that the culture we encounter today is largely composed of memes produced by human minds of long ago. In any developmental process there is an interaction between the formative constituents and their environment: genes depend on environmental input (nutrition, etc.) in order to fulfil their role as preservers and replicators of biological information; memes depend on human minds and other external media for their preservation and replication. The development of human minds depends on a combination of these processes: their innate potential is the result of an interaction between genes and the physical environment, and that potential is fulfilled as a result of interaction with memes (the cultural environment).

Yet there seems to be a danger here of implying that cultural evolution is not after all a genuinely distinctive process. If minds are ultimately the product of interaction between genes and the environment (even though much of that environment is cultural), then this seems no different from any other Darwinian process. Fortunately for memetics this problem is illusory: rather, minds are the unique product of an interaction between two quite independent Darwinian processes, one biological and the other cultural. The first is responsible for the mind's innate potential, and the other for the realization of that potential. Memes form part of the mind's environment, but they are a part that is governed by an evolutionary algorithm.

The advantage of this view is that it provides a systematic algorithm for the study of cultural change, without undermining our view of humans as genuinely conscious, creative and intentional beings. Minds do what we have always known that they do: they think, they are aware, they create and they feel. These processes provide the mechanisms of replication, variation and selection on which memetic evolution depends – but the fact that conscious minds provide those mechanisms does not contradict the claim that the consequent evolution is truly Darwinian. Cultural information is copied, varies and is selected, which is all that is needed for evolution to occur. The consciousness that is involved (at least some of the time) in these mechanisms serves no more to undermine the unconsciousness of the cultural evolutionary algorithm than the emotions and awareness that are involved in human sexual reproduction serve to undermine the unconsciousness of the biological evolutionary algorithm. Both cases exemplify the mundane scientific fact that we can often provide different levels of description (chemical, biological, psychological, cultural, etc.) for the same process.

Variation and Selection

At one level, then, alterations and novelties in cultural information, and their differential survival rates, can be described as processes involving human creativity and decision making. At another, though, they can be seen as memetic processes of variation and selection. In common with genes, memes can be shown to vary via both mutation and recombination, and from this perspective the shifting patterns of cultural change can be understood more clearly. Like genes, some portions of cultural material will be more likely than others to mutate; some directions of mutation will be more likely than others; and the ways in which cultural information is able to exert its effects on the world will also influence the changes that can take place. Whether the result of mutation or recombination, the variations that arise will be random with respect to fitness. Of course cultural evolution is played out at a far higher pace than biological evolution, and in culture many unsuitable variations are discarded so quickly that nobody other than their originator is aware of them. Nonetheless, the vision of cultural change as determined by recombination and random mutation can give us some insight into the nature of that change.

It should lead us to expect, for example, that the direction of cultural evolution will be largely dependent on what has gone before: both recombination and mutation are processes by which *existing* information is altered. This is in sharp contrast to the view of science, for example, as being directed primarily by an inexorable movement *towards* the truth; rather, the primary influence on its direction comes from behind, in the form of existing observations and theories. Similarly, we should expect to see the selection of information with the best fit to its environment, not in absolute terms but *relative to* what other information is available. Just as in genetics, we should not expect a certain variation automatically to appear because it would be useful or is an improvement on what has gone before. Like biology, culture has to wait upon the variations that are actually thrown up against the background of what already exists – and this is consistent with what history teaches us about many (to us) outlandish theories and perspectives that seemed reasonable and valid to our forebears.

Indeed, selection will often depend on a novelty's compatibility with the rest of the meme pool. In their bid to gain and retain our attention, memes will succeed best if they fit in with facts and skills that we have already absorbed, being influenced particularly by those to which we are greatly attached. Other factors in the selection process will include the best available evidence, the physical environment, and the dictates of human psychology. As cultural material accumulates, then, its influence on future variations will increase, and this explains why lateral thinking and novel theories are often much more welcome in emerging cultural areas than in established ones. The differential success of cultural variations will usually be determined more by the environment (memetic, genetic and physical) than by their own content.

The Selfish Meme: A Critical Reassessment

Richard Dawkins describes the essence of his selfish gene theory as the insight "that there are two ways of looking at natural selection, the gene's angle and that of the individual", and adds that such a change of vision can, at its best, "usher in a whole climate of thinking, in which many exciting and testable theories are born, and unimagined facts laid bare".[2]

What, then, is the essence of his selfish meme hypothesis? It is the insight that there are two ways of looking at cultural change, the meme's angle and that of the human individual. We can see culture's development as the result of human aspirations, creativity, intellect and effort – or we can see it as the product of memetic evolution. These are, as Dawkins puts it, "two views of the same truth".[3]

What theories are given birth, which facts laid bare by this change of vision? One of the most significant implications of the theory of the selfish *gene* is that the individual organism was not an inevitable outcome of genetic evolution: it so happens that genes have banded together to build survival machines, but the only crucial feature of any form of evolution is the replicator – the unit of selection. Although organisms clearly exist, and have a perspective from which the world of genes is irrelevant to their everyday lives, fundamentally their lives and evolution are determined by that world.

No analogous insight arises from the theory of the selfish *meme*, because memes do not build survival machines. Their replicative mechanisms, and the means of their variation and selection, lie in genetically determined human faculties, not in vehicles that they themselves build. The familiarity of the individual organism lent wonder to the claims that Dawkins made for the power of the selfish gene, but memes are not bundled up in a comparable way: analogously to the early genes, they are peppered freely throughout the cultural environment.

Memeticists like Daniel Dennett and Susan Blackmore, who disagree with this claim, have concluded that we are meme machines as much as gene machines, and are led to the assumption that "there is no conscious self inside" those machines; that "a complex interplay of replicators and environment" is all there is to life.[4] They are mistaken: it is not *all there is* to life, but merely *one way of describing* life. The other way – in terms of intellect and consciousness, desires and hopes, beliefs and emotions – is equally valid.

Nonetheless, there *is* an important insight to be gained by taking the perspective of the selfish meme: that cultural evolution is an autonomous process over which we exercise a limited amount of control. Our sense of self is not illusory, but our sense of control over the collective products of our minds may well be. Although our minds provide the mechanisms of memetic evolution, there is a very real sense in which the directions of that evolution are independent of us. As individuals we develop new ideas, give responses to existing cultural material and make an undeniable impact on the world around us – but we can do little about the ideas, responses and impact of other people, and each of us is so heavily outnumbered by everyone else that our own contributions are bound to be limited. The existence, moreover, of such a vast body of cultural material means that the success or failure of novel hypotheses, technological inventions or even ethical opinions will be determined more by their relative fitness for this immense meme pool than by their intrinsic merits.

Yet even this should not lead to despair. Although we can do nothing about the soil on which our cultural seeds fall, still we should care about the type of seeds that we plant. Scientists do not give up on their search for the truth just because there is a chance that they are starting from a position bogged down by the legacy of past theses rather than supported on the shoulders of the giants who originated them. The politics of funding, the stature of individual scientists and even the language that they speak may all influence the chances of their own research bearing fruit, but this need not detract from the value of the enterprise in which they are engaged. The same applies in any other area of culture, too. In religion, for instance, there are myriad different points of view, and their differential success (in terms of numbers of people who subscribe to each) may be affected by factors unrelated to their validity – but again this does not mean that there is no truth to be found, and no value in the search for it.

From the perspective of the selfish meme we can see that culture's development will ultimately be determined by a complex interplay between memes and their environment. The content of those memes, however, is our responsibility.

Acknowledgements

The work towards my doctoral thesis, on which this book was originally based, was not only supervised but also constantly inspired and encouraged by George Botterill. His generous gifts of time and thought were greatly appreciated. My thanks are also due to Peter Carruthers for his help and guidance during that period.

I am grateful for the support of the late Terence Moore and of Stephanie Achard at the Cambridge University Press and for the constructive comments, on earlier drafts, of two anonymous reviewers.

No work towards rewriting this book would have been possible without the provision of safe and enjoyable childcare for my sons, for a few precious hours each week. For this I am especially grateful to Jocelyn Lewry, to the Nursery and Kindergarten staff at The Croft School and Sally Jones in particular, and for a brief period to the staff at The Ark Nursery. Equally sustaining in this regard have been the times that the boys have spent with their Grandmas.

Encouragement for my endeavours has been unstintingly provided by my family and friends – particularly my parents – and for this I am very thankful.

Most of all, my gratitude is due to my husband, Keith, whose support has been immeasurable. From unforgivingly precise proofreader to most thought-provoking critic; from grand fixer of all computer problems to chief provider of practical examples; from most generous financial supporter to principal retriever of philosopher's head from the clouds; from most effective dismisser of self-doubt to foremost reinforcer of faith . . . none of this would have been possible without him.

Now to him who is able to do immeasurably more than all we ask or imagine, according to his power that is at work within us, to him be glory in the church and in Christ Jesus throughout all generations, for ever and ever! (Ephesians 3: 20–1)

Notes

Chapter 1 Introduction

1. Dawkins 1989: 189.
2. Jeremy H. Barkhow 1989: 118.

Chapter 2 The Meme Hypothesis

1. Dawkins 1989: 194.
2. Ibid. 191.
3. Ibid. 200.
4. Dawkins 1982: 290.
5. Dawkins 1989: 193.
6. Dawkins 1982: 109.
7. Ibid.
8. Anthony Flew ed. 1979: 353.
9. C. Darden and J. A. Cain 1989: 109.
10. Wilson, "Heredity," in Michael Ruse 1989: 247.
11. Elliott Sober 1993: 191.
12. R. Boyd and P. J. Richerson 1985: 13.

Chapter 3 Cultural DNA

1. This theory has been defended by Fred Dretske, 1988, amongst others.
2. The following discussion draws heavily on the work of Nicholas Agar, 1993. Agar's thesis was itself in part a response to Fodor, 1990.
3. Agar, 1993: 1.
4. An example taken from the work of Kandel et al. on *Aplysia californica.* See for instance Eric R. Kandel and James H. Schwartz eds. 1981, ch. 52.

Chapter 4 The Replication of Complex Culture

1. Blackmore 1999
2. Quoted variously by Arthur Koestler and others. Originally Simon 1962.
3. Richard Dawkins 1982: 251.
4. Albert R. Meyer and Eric Lehman 2002.
5. Dawkins 1976.
6. Dawkins 1982: 251. He is quoting from Dawkins 1976.
7. Koestler 1978.
8. David L. Hull, in Robert Aunger ed. 2000: 55.

Chapter 5 Variation

1. Richard Dawkins 1986: 307.
2. Ibid.
3. Liane Gabora 1997: 4.1.
4. Cf. George Botterill 1995.
5. Gabora 1997 – though notice that the original point of her quotation was not to support the views that I express here.
6. Sperber 1996: 29.
7. Ibid. 118.
8. Ibid. 29.
9. Pinker 1994.
10. Dennett 1995: 355.
11. Ibid. 354.

Chapter 6 Selection

1. Dan Sperber 1996: 54.
2. Liane Gabora 1997: 3.2.
3. Mario Vaneechootte 1998.
4. Rosaria Conte in Robert Aunger ed. 2000: 103.
5. See Rowan Bayne, Ian Horton, Tony Merry, Elizabeth Noyes et al. 1999: 62–3 for further discussion of this point.
6. Sperber 1996: 71ff.
7. Kevin Laland and John Odling-Smee in Aunger ed. 2000: 134.
8. P. E. Griffiths 1993.

Chapter 8 The Human Mind: Meme Complex with a Virus?

1. See, e.g., M. Midgely 1979, a surprisingly vitriolic attack in which both genes and memes are characterized as conscious; John Bowker 1995, ch. 8, an equally passionate rebuttal of meme theory, which is based on a different series of misconceptions.
2. E.g., Dawkins 1993a and 1993b.
3. Dawkins 1993b.
4. Ibid.

5. Ibid.
6. Dawkins 1993a: 26.
7. Shrader 1980: 281.
8. Dawkins 1989: 36.
9. Ibid. 199.
10. Dennett 1991: 210.
11. This interpretation has been taken from Dennett 1990, and from the relevant parts of Dennett 1991. Dennett 1995 is also relevant, but (on memes, anyway) does not add much to Dennett 1991.
12. Dennett 1991: 200 (emphases Dennett's).
13. Ibid. 207.
14. Dennett 1995: 349.
15. Ibid. 347–8.
16. Dennett 1991: 203–4.
17. Blackmore 1999: 17.
18. Conte in Robert Aunger ed. 2000: 113.
19. Dawkins 1982: 20.
20. 1982.
21. Dawkins 1982: 117.
22. Ibid. 292.
23. Ibid. 210.
24. Ibid. 263–4.
25. Dennett 1990: 128.
26. Dennett 1991: 416.
27. Dennett 1995: 366.
28. Ibid.
29. Dawkins 1989: 235.
30. Elliott Sober 1993, ch. 7.
31. Andy Clark 1995: 23.

Chapter 9 The Meme's Eye View

1. Blackmore 1999: 65.
2. Blackmore in Robert Aunger ed. 2000: 32.
3. Ibid. 36.
4. Ibid. 37.
5. Blackmore 1999: 192.
6. Ibid. 197.
7. Terence Deacon 1997.
8. Blackmore in Aunger ed. 2000: 26.
9. J. B. Sykes 1982: 231.
10. Blackmore in Aunger ed. 2000: 34.
11. Kevin Laland and John Odling-Smee in Aunger ed. 2000: 128.
12. Blackmore in Aunger ed. 2000: 27.
13. Plotkin in Aunger ed. 2000: 77.
14. Ibid. 76.
15. Ibid. 78.

16. Ibid. 79.
17. Ibid. 80.
18. Sperber in Aunger ed. 2000: 164.
19. Sperber 1996: 102.
20. Ibid.
21. Ibid. 106.
22. Sperber in Aunger ed. 2000: 171.
23. Ibid. 169.
24. Dawkins: foreword to Susan Blackmore 1999: vii–xvii.
25. Sperber 1996: 101.
26. Sperber in Aunger ed. 2000: 172.
27. Ibid.
28. Sperber 1996: 106.
29. Boyd and Richerson in Aunger ed. 2000: 155.
30. Ibid.
31. Ibid. 159.
32. Mario Vaneechootte 1998.

Chapter 10 Early Cultural Evolution

1. Richard Dawkins 1989: 14ff.
2. Ibid. 19.
3. Ibid.
4. Ibid.
5. Byrne and Russon 1998.
6. Ibid. 13. This quotation was underlined in the original.
7. Deacon 1997, ch. 4.
8. R. Leakey and R. Lewin 1992: 302–3.
9. Cf. ibid. 298.
10. Ibid. 300.
11. Frans de Waal (primatologist), quoted in Leakey and Lewin, ibid. 300–1.
12. George Botterill 1994. His reference is to A. Whiten (1993) "Evolving a theory of mind: the nature of non-verbal mentalism in other primates", in Baron-Cohen, Tager-Flusberg and Cohen eds. 1993, *Understanding Other Minds.*
13. G. Ettlinger 1987: 173.
14. A. N. Meltzoff and R. W. Borton 1979.
15. V. W. Gunderson 1983.
16. Meltzoff and Borton 1979: 404.
17. Leakey and Lewin 1992: 289–90.
18. A. H. Brodick 1960: 85.
19. Dawkins 1986: 190.
20. Leakey and Lewin 1992: 170.
21. Ibid. 172.
22. National Association for Gifted Children 1998: 4.
23. Byrne and Russon 1998.
24. Ibid. 22.
25. Deacon 1997: 94.

Chapter 11 Memetic DNA

1. Dawkins 1986: 217–18.
2. See Sleven Pinker 1994: 106ff for a discussion of X-bar theory, which claims that all phrases in all languages are governed by one plan – so the rules are *all* that matter.
3. Deacon 1997, ch. 3.
4. Ibid. 92–3 (my italics).
5. Ibid. ch. 4.
6. Ibid. 146.
7. Elliott Sober 1993: 41.
8. Mike Darton and John O. E. Clark 1994: 308.
9. Robert Aunger 2002.
10. Ibid. 154.
11. Ibid. 157.
12. Ibid. 154.
13. Ibid. 157.
14. Ibid. 154.
15. Ibid.
16. Dawkins 1989: 33–4.
17. Ibid. 196.
18. Bloch in Robert Aunger ed. 2000: 194.
19. Ibid. 200.
20. Ibid. 199.
21. Ibid. 201.
22. David L. Hull in Aunger ed. 2000: 45.
23. Chomsky in Richard L. Gregory 1987: 420.
24. Darton and Clark 1994: 524.
25. Chomsky in Richard L. Gregory 1987: 420.
26. Ibid. 421.
27. Ibid. 420.

Chapter 12 Memes and the Mind

1. Susan Blackmore in Robert Aunger ed. 2000: 29.
2. Blackmore 1999: 233.
3. Ibid.
4. Conte in Aunger 2000: 87 (original in italics).
5. Aunger in Aunger ed. 2000: 14 – here Aunger is summarizing Conte.
6. Sheila Dainow 1995.
7. Matthew 13: 1–9 and 18–23.
8. The remainder of this chapter is taken largely from an earlier paper which I wrote jointly with Keith Distin: K. E. Distin and K. W. Distin 1996.
9. Wallace 1989: 1.
10. Ibid. 10.
11. Ibid. 11.
12. Rose 1999a.

13. Rose 1999b.
14. David L. Hull 1999.
15. Richard Dawkins 1989: 215.

Chapter 13 Science, Religion and Society: What Can Memes Tell Us?

1. Douglas Shrader 1980.
2. Richard Dawkins 1993b.
3. Russell 1927: 19.
4. Ibid. 20.
5. Michael Poole ed. 1997: 5.
6. Dawkins 1993a: 25.
7. Ibid.
8. Smith 1976.
9. Lewis 1941: 17–21.

Chapter 14 Conclusions

1. National Association for Gifted Children – Education and Research Sub-Committee 1998: 4.
2. Dawkins 1989: viii–ix.
3. Ibid. ix – though notice that Dawkins's use of this phrase did not refer to memes, and was not intended to support the view of memes that I express here.
4. Blackmore 1999: 241–6.

Bibliography

Agar, Nicholas (1993) What do frogs really believe? *Australasian Journal of Philosophy* 71:1–12.

Alp, R. (1993) Meat eating and ant dipping by wild chimpanzees in Sierra Leone, *Primates* 34:463–8.

Andreski, S., ed. (1971) *Herbert Spencer*, Nelson.

Antilla, R. (1972) *An introduction to historical and comparative linguistics*, Macmillan.

Aunger, Robert (1999) Culture vultures, *The Sciences* 39 (5):36–42.

Aunger, Robert, ed. (2000) *Darwinizing culture – the status of memetics as a science*, Oxford University Press.

Aunger, Robert (2002) *The electric meme: a new theory of how we think*, Free Press.

Avers, C. J. (1989) *Process and pattern in evolution*, Oxford University Press.

Ball, J. A. (1984) Memes as replicators, *Ethology and Sociobiology* 5:145–61.

Barkhow, Jeremy H. (1989) The elastic between genes and culture, *Ethology and Sociobiology* 10:111–26.

Barnett, S. M., ed. (1962) *A century of Darwinism*, Mercury.

Bayne, Rowan; Horton, Ian; Merry, Tony; Noyes, Elizabeth; & McMahon, Gladeana (1999) *The counsellor's handbook – a practical A-Z guide to professional and clinical practice*, Stanley Thorne.

Beer, Francis A. (1999) Memetic meanings – a commentary on Rose's paper: Controversies in meme theory, *Journal of Memetics – Evolutionary Models of Information Transmission* [Internet] 3 (1), June 1999. Available from: <http://jom-emit.cfpm.org/1999/vol3/beer_fa.html>

Benzon, William L. (2002) Colorless green homunculi, *Human Nature Review* 2:454–62.

Bickerton, Derek (1990) *Language and species*, University of Chicago Press.

Blackmore, Susan (1999) *The meme machine*, Oxford University Press.

Blackmore, Susan (2000) The power of memes, *Scientific American* 283 (4):52–61.

Bonjour, Laurence (1985) *The structure of empirical knowledge*, Harvard University Press.

Botterill, George (1993) "Functions and functional explanation". Unpublished manuscript.

Botterill, George (1994) Personal communication.

Botterill, George (1995) "How can we learn anything from thought experiments?" Unpublished manuscript.

Botterill, George, & Carruthers, P. (1999) *The philosophy of psychology*, Cambridge University Press.

Bowker, John (1995) *Is God a virus? Genes, culture and religion*, SPCK.

Boyd, Gary (2001) The human agency of meme machines, An extended review of "The meme machine" by Susan Blackmore, *Journal of Memetics – Evolutionary Models of Information Transmission* [Internet] 5 (1), March 2001. Available from: <http://jom-emit.cfpm.org/2001/vol5/boyd_g.html>

Boyd, R., & Richerson, P. J. (1985) *Culture and the evolutionary process*, University of Chicago Press.

Brodick, A. H. (1960) *Man and his ancestry*, Scientific Book Club.

Bryson, Bill. (1990) *Mother tongue*, Penguin Books.

Burhoe, Ralph Wendell (1979) Religion's role in human evolution: the missing link between ape-man's selfish genes and civilized altruism, *Zygon* 14: 135–62.

Byrne, Richard W., & Russon, Anne E. (1998) Learning by imitation: a hierarchical approach, *Behavioural and Brain Sciences* 21 (5):667–84.

Carruthers, Peter (1992) *Human knowledge and human nature*, Oxford University Press.

Carruthers, Peter (1994) "Prolegomena to a project: the natural involvement of language in thought". Unpublished manuscript.

Carruthers, Peter (1995) *Language, thought and consciousness*, Cambridge University Press.

Cavalli-Sforza, L. L., & Feldman, M. (1981) *Cultural transmission and evolution: a quantitative approach*, Princeton University Press.

Christians in Science, *Proceedings of the Conference on Creation and Evolution*, 1993, Regent's College London (1993), conference chairman Oliver Barclay.

Clark, Andy (1995) Language: the ultimate artifact, unpublished first draft of chapter 10 in Andy Clark (1997) *Being there: putting brain, body and world together again*, MIT Press.

Clark, S. R. L. (1993) Minds, memes and rhetoric, *Inquiry* 36:3–16.

Cloak, F. T. (1975) Is a cultural ethology possible? *Human Ecology* 3:161–82.

Critchlow, P. (1982) *Mastering chemistry*, Macmillan.

Dahlbom, B., ed. (1993) *Dennett and his critics*, Blackwell.

Dainow, Sheila (1995) Uncovering a TA script with pictures, *Counselling* 6: 291–3.

Darden, C., & Cain, J. A. (1989) Selection type theories, *Philosophy of Science* 56: 106–29.

Darton, Mike, & Clark, John O. E. (1994) *The Dent dictionary of measurement*, J. M. Dent.

Darwin, Charles (1859) *The origin of species*, Penguin 1968 edition.

Darwin, Charles (1887) Autobiographical chapter in Darwin, F., ed., *Life and letters*, vols. 1–3, John Murray.

Dawkins, Richard (1976) Hierarchical organisation: a candidate principle for ethology, in P. P. G. Bateson & R. A. Hinde, eds (1976) *Growing points in ethology*, Cambridge Universty Press, pp. 7–54.

Dawkins, Richard (1978) Reply to Fix and Greene, *Contemporary Sociobiology* 7.

Dawkins, Richard (1982) *The extended phenotype*, Oxford University Press.

Dawkins, Richard (1986) *The blind watchmaker*, Penguin Books.

Dawkins, Richard (1989) *The selfish gene*, revised edition, Oxford University Press.

Dawkins, Richard (1993a) Viruses of the mind, in B. Dahlbom, ed. (1993) *Dennett and his critics*, Blackwell, pp. 13–27.

Dawkins, Richard (1993b) Is religion just a disease? *Daily Telegraph*, 15 December, p. 18.

Dawkins, Richard (1995a) A reply to Poole, *Science and Christian Belief* 7:45–50.

Dawkins, Richard (1995b) *River out of Eden*, Weidenfeld and Nicolson.

Deacon, Terence (1997) *The symbolic species – the co-evolution of language and the human brain*, Allen Lane, The Penguin Press.

Delfgaauw, Bernard (1961) *The theory of Teilhard de Chardin*, Collins (Fontana).

Dennett, Daniel C. (1990) Memes and the exploitation of imagination, *Journal of Aesthetics and Art Criticism* 48:127–35.

Dennett, Daniel C. (1991) *Consciousness explained*, Penguin Books.

Dennett, Daniel C. (1993) Living on the edge, *Inquiry* 36:135–8.

Dennett, Daniel C. (1995) *Darwin's dangerous idea – evolution and the meanings of life*, Allen Lane, Penguin Press.

Dennett, Daniel C. (1999) *The evolution of culture*, Charles Simonyi Lecture at Oxford University, February 17 1999 [Internet] Available from:<http://www.edge.org/3rd_culture/dennett/dennett_p1.html>.

Desmond, A., & Moore, J. (1991) *Darwin*, Michael Joseph.

Distin, K. E. (1997) Cultural evolution – the meme hypothesis, in *Proceedings of the Conference on Evolution, or How did we get here?* 1997, Westminster College Oxford, Christ and the Cosmos.

Distin, K. E. (1999) : Interfaith issues in Religious Education: a response to Lat Blaylock, *REsource* 21 (3):16–18.

Distin, K. E., & Distin, K. W. (1996) Human design methods, in *Proceedings of the Conference on Design in Nature?*, Birmingham 1996, conference chairman Colin Russell, Christians in Science.

Donald, Merlin (1993) Reply to Plotkin, *Behavioural and Brain Science* 16:782–3.

Dretske, F. (1988) *Explaining behaviour*, MIT Press.

Durham, W. H. (1990) Advances in evolutionary culture theory, *Annual Review of Anthropology* 19:187–210.

Durkheim, Émile (1938) *The rules of sociological method*, University of Chicago Press.

Ettlinger, G. (1987) Cross-modal sensory integration, in Richard L. Gregory, ed. (1987) *The Oxford companion to the mind*, Oxford University Press, pp. 173–4.

Flew, Anthony, ed. (1979) *Dictionary of philosophy*, Pan Books.

Fodor, J. A. (1987) *Psychosemantics*, MIT Press.

Fodor, J. A. (1990) *A theory of content and other essays*, MIT Press.

Fregmen, A., & Fay, D. (1980) Over-extensions in production and comprehension: methodological clarification, *Journal of Child Language* 7:205–11.

Gabora, Liane (1997) The origin and evolution of culture and creativity, *Journal of Memetics – Evolutionary Models of Information Transmission* [Internet] 1 (1), May 1997. Available from <http://jom- emit.cfpm.org/1997/vol1/gabora_l.html>

Gabora, Liane (1999) A review of "The Meme Machine" by Susan Blackmore, *The Journal of Artificial Societies and Social Simulation* [Internet] 2 (2), March 1999. Available from: <http://jasss.soc. surrey.ac.uk/2/2/review2.html>

Gatherer, Derek (1999) A plea for methodological Darwinism – a commentary on Rose's paper: Controversies in meme theory, *Journal of Memetics – Evolutionary Models of Information Transmission* [Internet] 3 (1), June 1999. Available from: <http://jom-emit. cfpm.org/1999/vol3/gatherer_dg2.html>

Goswami, Usha (1990) "Developmental psychology". Cambridge University NST 1B lecture notes.

Gregory, Richard L., ed. (1987) *The Oxford companion to the mind*, Oxford University Press.

Griffiths, P. E. (1993) Functional analysis and proper functions, *British Journal for the Philosophy of Science* 44:409–22.

Gross, John (1991) *The Oxford book of essays*, Oxford University Press.

Gunderson, V. W. (1983) Development of cross-modal recognition in infant pigtail monkeys (*Macaca nemestrina*), *Developmental Psychology* 19:389–404.

Hacking, I. (1983) *Representing and intervening*, Cambridge University Press.

Heylighen, Francis (1998) What makes a meme successful? Selection criteria for cultural evolution, in *Symposium on memetics*, 15th International Congress on Cybernetics, Namur (1998), symposium chairmen Francis Heylighen & Mario Vaneechoutte. Available from: <http://pespmc1.vub.ac.be/papers/MemeticsNamur.html>

Hofstadter, Richard (1944) *Social Darwinism in American thought*, 1959 edition, George Braziller.

Honderich, Ted, ed. (1995) *The Oxford companion to philosophy*, Oxford University Press.

Hull, David L. (1988) A mechanism and its metaphysics: an evolutionary account of the social and conceptual development of science, *Biology and Philosophy* 3:123–55.

Hull, David L. (1999) Strategies in meme theory – a commentary on Rose's paper: Controversies in meme theory, *Journal of Memetics – Evolutionary Models of Information Transmission* [Internet] 3 (1), June 1999. Available from <http://jom-emit.cfpm.org/1999/vol3/hull_ dl.html>

Jones, B. E. M. (1995) "Is there a role for syntax/parsing in NLP?" Unpublished manuscript.

Kandel, Eric R., & Schwartz, James H., eds. (1981) *Principles of neural science*, Edward Arnold.

Karmiloff-Smith, Annette (1992) *Beyond modularity – a developmental perspective on cognitive science*, MIT Press.

Kenny, A. (1989) *The metaphysics of mind*, Clarendon Press.

Kirk, Robert (1993) "The best set of tools"? Dennett's metaphors and the mind-body problem, *The Philosophical Quarterly* 43:335–43.

Koestler, Arthur (1964) *The act of creation*, Hutchinson.

Koestler, Arthur (1978) *Janus – a summing up*, Hutchinson.

Kolenda, K. (1979) Introduction in Gilbert Ryle, *On thinking*, Blackwell.

Leakey, R., & Lewin, R. (1992) *Origins reconsidered*, Little, Brown & Co.

Levi, Primo (1992) *The periodic table*, Abacus.

Lewis, C. S. (1941) Bulverism, in Walter Hooper ed. (1996) *Compelling reason*, Fount, pp. 17–21.

Lewis, D. (1983) *Philosophical papers*, vol. 1, Oxford University Press.

Line, Christina Personal communication.

Lukes, Steven (1975) *Émile Durkheim – his life and work: a historical and critical study*, Peregrine.

Lycan, W. ed. (1990) *Mind and cognition: a reader*, Blackwell.

Lynch, A. et al. (1989) A model of cultural evolution of chaffinch song derived with the meme concept, *American Naturalist* 133:634–53.

Lynch, A. et al. (1993) A population memetics approach to cultural evolution in chaffinch song, *American Naturalist* 141:597–620.

McCrone, J. (1990) *The ape that spoke*, Picador.

McGinn, C. (1989) *Mental content*, Blackwell.

Magee, Bryan (1975) *Popper*, Collins (Fontana).

Malthus, T. R. (1826) *An essay on the principle of population*, John Murray.

Mason, Kelby (1998) Thoughts as tools: the meme in Daniel Dennett's work, in *Symposium on memetics*, 15th International Congress on Cybernetics, Namur (1998), symposium chairmen Francis Heylighen & Mario Vaneechoutte. Available from: <http://pespmc1. vub.ac.be/Conf/MemePap/Mason.html>

Masters, R. D. (1970) Genes, language and evolution, *Semiotica* 2:295–320.

Medawar, Peter (1990): *The threat and the glory*, Oxford University Press.

Meltzoff, A. N., & Borton, R. W. (1979) Intermodal matching by human neonates, *Nature* 282:403–4.

Meyer, Albert R., & Lehman, Eric (2002) Random variables: Expectation, *Mathematics for Computer Science Handouts and Course Notes* Week 12 (Fall 2002). Available from <http://theory.lcs.mit.edu/classes/6.042/fall02/handouts/ #week12>

Midgley, M. (1979) Gene-juggling, *Philosophy* 54 (210):439–58.

Millikan, Ruth G. (1984) *Language, thought, and other biological categories*, MIT Press.

Millikan, Ruth G. (1990) Compare and contrast Dretske, Fodor, and Millikan on teleosemantics, *Philosophical Topics* 18 (2):151–61.

Modelski, George (1999) An evolutionary theory of culture? – a commentary on Rose's paper: Controversies in meme theory, *Journal of Memetics – Evolutionary Models of Information Transmission* [Internet] 3 (1), June 1999. Available from: <http://jom-emit.cfpm.org/1999/vol3/modelski_g.html>

Morris, Desmond (1968) *The naked ape*, Corgi.

National Association for Gifted Children, Education and Research Sub-Committee (1998) *Help with bright children*, NAGC.

Oldroyd, D. R. (1983) *Darwinian impacts*, Open University Press.

Papineau, D. (1993) "The teleological theory of representation". Seminar delivered to Sheffield University Philosophy Department.

Pinker, Steven (1994) *The language instinct – the new science of language and mind*, Penguin Books.
Plotkin, H. C. (1993) Hunting memes, *Behavioural and Brain Science* 16:768–9.
Poole, Michael (1994) A critique of aspects of the philosophy and theology of Richard Dawkins, *Science and Christian Belief* 6:41–59.
Poole, Michael (1995) A response to Dawkins, *Science and Christian Belief* 7:51–8.
Poole, Michael, ed. (1997) *God and the scientists*, CPO – Design and Print.
Popper, Karl R. (1972) *Objective knowledge – an evolutionary approach*, Oxford University Press.
Porter, Roy (1987) *Man masters nature*, BBC.
Price, If (1999) Steps towards the memetic self – a commentary on Rose's paper: Controversies in meme theory, *Journal of Memetics – Evolutionary Models of Information Transmission* [Internet] 3 (1), June 1999. Available from: <http://jom-emit.cfpm.org/1999/vol3/price_if.html>
Research Scientists' Christian Fellowship, *Proceedings of the Conference on Evolution*, 1974, Bedford College London (1974), conference chairman Duncan Vere.
Robbins, T. W., & Cooper, P. J. (1988) *Psychology for medicine*, Edward Arnold.
Roberts, K. (1998) The academic and emotional needs of gifted children – a personal case study, *Gifted and Talented* 2:32–8.
Rose, Nick (1998) Controversies in meme theory, *Journal of Memetics – Evolutionary Models of Information Transmission* [Internet] 2 (1), June 1998. Available from: <http://jom-emit.cfpm.org/1998/vol2/rose_n.html>
Rose, Nick (1999a) Rationale for commentary on Rose's paper: Controversies in meme theory, *Journal of Memetics – Evolutionary Models of Information Transmission* [Internet] 3 (1), June 1999. Available from <http://jom-emit.cfpm.org/1999/vol3/rose_n_case_for_commentary.html>
Rose, Nick (1999b) Okay, but exactly "who" would escape the tyranny of the replicators? – reply to the commentaries on the author's paper: Controversies in meme theory, *Journal of Memetics – Evolutionary Models of Information Transmission* [Internet] 3 (1), June 1999. Available from: <http://jom-emit.cfpm.org/1999/vol3/rose_n.html>
Ruse, Michael (1982) *Darwinism defended*, Addison-Wesley.
Ruse, Michael, ed. (1989) *Philosophy of biology*, Macmillan.
Russell, Bertrand (1927) Why I am not a Christian, in Paul Edwards, ed. (1957) *Why I am not a Christian*, 1996 edition, Routledge, pp. 13–26.
Ryle, Gilbert (1963) *The concept of mind*, Penguin Books.
Ryle, Gilbert (1979) *On thinking*, Blackwell.
Schull, J. (1990) Are species intelligent? *Behavioural and Brain Sciences* 13:68–113.
Searle, John (1989) *Minds, brains and science*, Pelican.
Shafto, Michael, ed. (1985) *How we know*, Harper and Row.
Shrader, Douglas (1980) The evolutionary development of science, *Review of Metaphysics* 34:273–96.
Simon, Herbert (1962) The architecture of complexity, *Proceedings of the American Philosophical Society* 106:6, December 1962, pp. 467–82.
Singer, Peter ed. (1986) *Applied ethics*, Oxford University Press.
Sluckin, W. (1954) *Minds and machines*, Pelican.

Smith, George (1976) *How to defend atheism*, Speech delivered to the Society of Separationists [Internet]. Available from: <http://www. infidels.org/library/modern/george_smith/defending.html>

Sober, Elliott (1993) *Philosophy of biology*, Oxford University Press.

Spencer, Herbert (1862) *First principles*, 1900 edition, Williams & Norgate.

Sperber, Dan (1996) *Explaining culture – a naturalistic approach*, Blackwell.

Sperber, Dan (2000) Metarepresentations in an evolutionary perspective, in Dan Sperber, ed., *Metarepresentations: a multidisciplinary approach*, Oxford University Press, pp. 117–37.

Sperber, Dan, & Wilson, Deirdre (1986) *Relevance: communication and cognition*, Blackwell.

Steen, E. B. (1971) *Dictionary of biology*, Barnes and Noble.

Steinbeck, John (1979) *A life in letters*, Pan Books.

Sykes, J. B. (1982) *The concise Oxford dictionary*, seventh edition, Oxford University Press.

Tomasello, M. & Farrar, M. J. (1986) Object permanence and relational words: a training study, *Journal of Child Language* 13:495–505.

Vaneechoutte, Mario (1998) The replicator: a misnomer. Conceptual implications for genetics and memetics, in *Symposium on Memetics*, 15th International Congress on Cybernetics, Namur (1998), symposium chairmen Francis Heylighen & Mario Vaneechoutte. Available from <http://pespmc1.vub.ac.be/Conf/MemePap/Vaneechoutte.html>

Wallace, A. R. (1895) *Natural selection and tropical nature*, Macmillan.

Wallace, K. M. (1989) *An introduction to the design process*, Cambridge University Engineering Department.

Wilkins, John (1999): On choosing to evolve: strategies without a strategist – a commentary on Rose's paper: Controversies in meme theory, *Journal of Memetics – Evolutionary Models of Information Transmission* [Internet] 3 (1), June 1999. Available from: <http://jom-emit.cfpm.org/1999/vol3/wilkins_j2.html>

Young, R. M. (1985) *Darwin's metaphor*, Cambridge University Press.

Index